Review of NASA's Biomedical Research Program

Committee on Space Biology and Medicine
Space Studies Board
Commission on Physical Sciences, Mathematics, and Applications
National Research Council

NATIONAL ACADEMY PRESS
Washington, D.C.

NOTICE: The project that is the subject of this report was approved by the Governing Board of the National Research Council, whose members are drawn from the councils of the National Academy of Sciences, the National Academy of Engineering, and the Institute of Medicine. The members of the committee responsible for the report were chosen for their special competences and with regard for appropriate balance.

Support for this project was provided by Contract NASW 96013 between the National Academy of Sciences and the National Aeronautics and Space Administration. Any opinions, findings, conclusions, or recommendations expressed in this report are those of the authors and do not necessarily reflect the views of the sponsor.

Copyright 2000 by the National Academy of Sciences. All rights reserved.

Copies of this report are available from:

Space Studies Board
National Research Council
2101 Constitution Avenue, N.W.
Washington, D.C. 20418

Printed in the United States of America

THE NATIONAL ACADEMIES

National Academy of Sciences
National Academy of Engineering
Institute of Medicine
National Research Council

The **National Academy of Sciences** is a private, nonprofit, self-perpetuating society of distinguished scholars engaged in scientific and engineering research, dedicated to the furtherance of science and technology and to their use for the general welfare. Upon the authority of the charter granted to it by the Congress in 1863, the Academy has a mandate that requires it to advise the federal government on scientific and technical matters. Dr. Bruce M. Alberts is president of the National Academy of Sciences.

The **National Academy of Engineering** was established in 1964, under the charter of the National Academy of Sciences, as a parallel organization of outstanding engineers. It is autonomous in its administration and in the selection of its members, sharing with the National Academy of Sciences the responsibility for advising the federal government. The National Academy of Engineering also sponsors engineering programs aimed at meeting national needs, encourages education and research, and recognizes the superior achievements of engineers. Dr. William A. Wulf is president of the National Academy of Engineering.

The **Institute of Medicine** was established in 1970 by the National Academy of Sciences to secure the services of eminent members of appropriate professions in the examination of policy matters pertaining to the health of the public. The Institute acts under the responsibility given to the National Academy of Sciences by its congressional charter to be an adviser to the federal government and, upon its own initiative, to identify issues of medical care, research, and education. Dr. Kenneth I. Shine is president of the Institute of Medicine.

The **National Research Council** was organized by the National Academy of Sciences in 1916 to associate the broad community of science and technology with the Academy's purposes of furthering knowledge and advising the federal government. Functioning in accordance with general policies determined by the Academy, the Council has become the principal operating agency of both the National Academy of Sciences and the National Academy of Engineering in providing services to the government, the public, and the scientific and engineering communities. The Council is administered jointly by both Academies and the Institute of Medicine. Dr. Bruce M. Alberts and Dr. William A. Wulf are chairman and vice chairman, respectively, of the National Research Council.

COMMITTEE ON SPACE BIOLOGY AND MEDICINE

MARY JANE OSBORN, University of Connecticut Health Center, *Chair*
NORMA M. ALLEWELL, Harvard University
JAY C. BUCKEY, JR., Dartmouth-Hitchcock Medical Center
LYNETTE JONES, Massachusetts Institute of Technology
ROBERT A. MARCUS, VA Palo Alto Health Care System
LAWRENCE A. PALINKAS, University of California at San Diego
KENNA D. PEUSNER, George Washington University Medical Center
STEVEN E. PFEIFFER, University of Connecticut Medical School
DANNY A. RILEY, Medical College of Wisconsin
RICHARD SETLOW, Brookhaven National Laboratory
GERALD SONNENFELD, Morehouse School of Medicine
T. PETER STEIN, University of Medicine and Dentistry of New Jersey
JUDITH L. SWAIN, Stanford University School of Medicine

Staff
SANDRA J. GRAHAM, Study Director
ANNE K. SIMMONS, Senior Program Assistant

SPACE STUDIES BOARD

CLAUDE R. CANIZARES, Massachusetts Institute of Technology, *Chair*
MARK R. ABBOTT, Oregon State University
FRAN BAGENAL, University of Colorado
DANIEL N. BAKER, University of Colorado
ROBERT E. CLELAND, University of Washington
MARILYN L. FOGEL, Carnegie Institution of Washington
BILL GREEN, former member, U.S. House of Representatives
JOHN H. HOPPS, JR., Morehouse College
CHRIS J. JOHANNSEN, Purdue University
ANDREW H. KNOLL,* Harvard University
RICHARD G. KRON, University of Chicago
JONATHAN I. LUNINE, University of Arizona
ROBERTA BALSTAD MILLER, Columbia University
GARY J. OLSEN, University of Illinois at Urbana-Champaign
MARY JANE OSBORN, University of Connecticut Health Center
GEORGE A. PAULIKAS, The Aerospace Corporation (retired)
JOYCE E. PENNER, University of Michigan
THOMAS A. PRINCE, California Institute of Technology
PEDRO L. RUSTAN, JR., U.S. Air Force (retired)
GEORGE L. SISCOE, Boston University
EUGENE B. SKOLNIKOFF, Massachusetts Institute of Technology
MITCHELL SOGIN, Marine Biological Laboratory
NORMAN E. THAGARD, Florida State University
ALAN M. TITLE, Lockheed Martin Advanced Technology Center
RAYMOND VISKANTA, Purdue University
PETER W. VOORHEES, Northwestern University
JOHN A. WOOD, Harvard-Smithsonian Center for Astrophysics

JOSEPH K. ALEXANDER, Director

---

*Former member.

COMMISSION ON PHYSICAL SCIENCES, MATHEMATICS, AND APPLICATIONS

PETER M. BANKS, Veridian ERIM International, Inc., *Co-chair*
W. CARL LINEBERGER, University of Colorado, *Co-chair*
WILLIAM F. BALLHAUS, JR., Lockheed Martin Corporation
SHIRLEY CHIANG, University of California at Davis
MARSHALL H. COHEN, California Institute of Technology
RONALD G. DOUGLAS, Texas A&M University
SAMUEL H. FULLER, Analog Devices, Inc.
JERRY P. GOLLUB, Haverford College
MICHAEL F. GOODCHILD, University of California at Santa Barbara
MARTHA P. HAYNES, Cornell University
WESLEY T. HUNTRESS, JR., Carnegie Institution of Washington
CAROL M. JANTZEN, Westinghouse Savannah River Company
PAUL G. KAMINSKI, Technovation, Inc.
KENNETH H. KELLER, University of Minnesota
JOHN R. KREICK, Sanders, a Lockheed Martin Company (retired)
MARSHA I. LESTER, University of Pennsylvania
DUSA M. McDUFF, State University of New York at Stony Brook
JANET L. NORWOOD, Former Commissioner, U.S. Bureau of Labor Statistics
M. ELISABETH PATÉ-CORNELL, Stanford University
NICHOLAS P. SAMIOS, Brookhaven National Laboratory
ROBERT J. SPINRAD, Xerox PARC (retired)

MYRON F. UMAN, Acting Executive Director

Preface

In 1998, the Committee on Space Biology and Medicine (CSBM) completed a comprehensive report, *A Strategy for Research in Space Biology and Medicine in the New Century* (National Academy Press, Washington, D.C., 1998), that reviewed the status of space life sciences research in all of the disciplines funded by the National Aeronautics and Space Administration's (NASA's) life sciences program and laid out a detailed strategy for research during the International Space Station era. In that report, numerous biomedical research questions related to astronaut health and safety were identified as critical to NASA's long-duration flight program. Shortly after the report's publication, NASA requested that CSBM assess the agency's entire current program in biomedical research, both intramural and extramural, in light of the recommendations of the *Strategy* report.

After a series of discussions with NASA's Life Sciences Division, the committee began reviewing NASA's entire biomedical research program in December 1998 in order to assess the degree to which the program seemed likely to meet research needs over the next 10 years. The research priorities given in the 1998 *Strategy* report were to be used as a point of departure when considering future needs and priorities. Specifically, the committee agreed to examine the relationship between intramural and extramural biomedical research activities sponsored by the agency and to review the content and program organization of both. The roles of the NASA Specialized Centers of Research and Training and the National Space Biomedical Research Institute, in the biomedical program, were also to be examined. The review was to cover all NASA biomedical research activities, including those currently conducted in conjunction with operational medical and aerospace medicine programs.

Some of the specific points the committee considered in developing its recommendations were the following:

- The balance of discipline areas emphasized in the current program;
- The degree to which studies of fundamental cellular and physiological mechanisms are addressed in each discipline program;
- The balance between ground and flight investigations;

- NASA plans for the development and validation of physiological and psychological countermeasures;
- Plans for epidemiology and monitoring;
- Plans for validation of animal models; and
- The extent to which programs are supporting new, advanced technologies and methodologies.

The committee made use of a variety of sources in gathering information for this study. Documents available to the committee included FY 1998 and FY 1999 life sciences budget information, the 1998 and 1999 Life Sciences Task Book, the first annual report of the National Space Biomedical Research Institute (NSBRI) and 1998 and 1999 program budget information, the Countermeasure Evaluation and Validation Project Plan, the International Space Station Medical Operations Requirements Document and relevant sections of the Astronaut Medical Evaluation Requirements Document, and NASA Research Announcements for 1998 and 1999. In addition, the *Proceedings of the First Biennial Biomedical Investigators' Workshop*, held in January 1999, provided valuable current information. In addition to receiving briefings from NASA and NSBRI spokespersons, the committee as a whole held one meeting at Johnson Space Center, and a subgroup visited Ames Research Center to learn about the activities at that site relevant to biomedical research. These visits provided a vast amount of useful information, and the committee wishes to express its considerable appreciation of the hard work that went into the centers' preparation for the visits and the thoroughness and candor of the briefings and discussions.

Acknowledgment of Reviewers

This report has been reviewed by individuals chosen for their diverse perspectives and technical expertise, in accordance with procedures approved by the National Research Council's (NRC's) Report Review Committee. The purpose of this independent review is to provide candid and critical comments that will assist the authors and the NRC in making the published report as sound as possible and to ensure that the report meets institutional standards for objectivity, evidence, and responsiveness to the study charge. The contents of the review comments and draft manuscript remain confidential to protect the integrity of the deliberative process. The committee wishes to thank the following individuals for their participation in the review of this report:

James Bagian, Environmental Protection Agency,
Norman Bell, Medical University of South Carolina,
Robert A. Greenes, Harvard Medical School,
Robert Langer, Massachusetts Institute of Technology,
Robert Nerem, Georgia Institute of Technology,
Gary Paige, University of Rochester, and
Edward Schultz, University of Wisconsin Medical School.

Although the individuals listed above have provided many constructive comments and suggestions, responsibility for the final content of this report rests solely with the authoring committee and the NRC.

Contents

EXECUTIVE SUMMARY 1

1 INTRODUCTION 7
    Reference, 10

2 SENSORIMOTOR INTEGRATION 11
    Introduction, 11
    NASA's Current Research Program in Sensorimotor Integration, 12
    Programmatic Balance, 15
        Balance of Subdiscipline Areas, 15
        Balance of Ground and Flight Investigations, 15
        Emphasis Given to Fundamental Mechanisms, 15
        Utilization and Validation of Animal Models, 15
    Development and Validation of Countermeasures, 16
    Epidemiology and Monitoring, 17
    Support of Advanced Technologies, 17
    Summary, 18
    Bibliography, 18

3 BONE PHYSIOLOGY 19
    Introduction, 19
    NASA's Current Research Program in Bone Physiology, 20
        Basic Research, 20
        Animal Studies, 21
        Human Studies, 21

Programmatic Balance, 22
    Balance of Subdiscipline Areas, 22
    Balance of Ground and Flight Investigations, 22
    Emphasis Given to Fundamental Mechanisms, 22
    Utilization and Validation of Animal Models, 22
Development and Validation of Countermeasures, 23
Epidemiology and Monitoring, 23
Support of Advanced Technologies, 24
Summary, 25
Bibliography, 25

4 MUSCLE PHYSIOLOGY — 26
Introduction, 26
NASA's Current Research Program in Muscle Physiology, 27
Programmatic Balance, 28
    Balance of Subdiscipline Areas, 28
    Balance of Ground and Flight Investigations, 28
    Emphasis Given to Fundamental Mechanisms, 28
    Utilization and Validation of Ground and Animal Models, 29
Development and Validation of Countermeasures, 29
Epidemiology and Monitoring, 30
    Plans for Monitoring Crew Health and Fitness on the International Space Station, 30
Support of Advanced Technologies, 31
Summary, 31
References, 32

5 CARDIOVASCULAR AND PULMONARY SYSTEMS — 33
Introduction, 33
NASA's Current Research Program in Cardiovascular and Pulmonary Systems, 34
Programmatic Balance, 35
    Balance of Subdiscipline Areas, 35
    Balance of Ground and Flight Investigations, 35
    Emphasis Given to Fundamental Mechanisms, 36
    Utilization and Validation of Animal Models, 36
Development and Validation of Countermeasures, 36
Epidemiology and Monitoring, 37
    Orthostatic Intolerance, 38
    Cardiac Atrophy, 38
    Arrhythmias, 38
    Pulmonary, 38
Support of Advanced Technologies, 39
Summary, 39
Bibliography, 39

6 ENDOCRINOLOGY AND NUTRITION — 40
Introduction, 40
NASA's Current Research Program in Endocrinology and Nutrition, 40
Programmatic Balance, 42
   Balance of Subdiscipline Areas, 42
   Balance of Ground and Flight Investigations, 43
   Emphasis Given to Fundamental Mechanisms, 43
   Utilization and Validation of Ground and Animal Models, 43
Development and Validation of Countermeasures, 44
Epidemiology and Monitoring, 45
Support of Advanced Technologies, 45
Summary, 45
References, 45

7 IMMUNOLOGY AND MICROBIOLOGY — 46
Introduction, 46
NASA's Current Research Program in Immunology and Microbiology, 47
Programmatic Balance, 48
   Balance of Subdiscipline Areas, 48
   Balance of Ground and Flight Studies, 48
   Emphasis Given to Fundamental Mechanisms, 48
   Utilization and Validation of Animal Models, 49
Development and Validation of Countermeasures, 49
Epidemiology and Monitoring, 49
Support of Advanced Technologies, 50
Summary, 50
References, 50

8 RADIATION BIOLOGY — 51
Introduction, 51
NASA's Current Research Program in Radiation Biology, 52
Programmatic Balance, 54
   Balance of Subdiscipline Areas, 54
   Balance of Ground and Flight Investigations, 55
   Emphasis Given to Fundamental Mechanisms, 55
   Utilization and Validation of Animal Models, 55
Development and Validation of Countermeasures, 56
Epidemiology and Monitoring, 56
Support of Advanced Technologies, 56
Summary, 57
References, 57

9 BEHAVIOR AND PERFORMANCE — 58
Introduction, 58
NASA's Current Research Program in Behavior and Performance, 59

Programmatic Balance, 62
  Balance of Subdiscipline Areas, 62
  Balance of Ground and Flight Investigations, 62
  Emphasis Given to Fundamental Mechanisms, 63
  Utilization and Validation of Animal Models, 63
Development and Validation of Countermeasures, 64
Epidemiology and Monitoring, 66
Support of Advanced Technologies, 66
Summary, 67
References, 67

10 SETTING PRIORITIES IN RESEARCH 68
Loss of Weight-bearing Bone and Muscle, 68
Vestibular Function, the Vestibulo-ocular Reflex, and Sensorimotor Integration, 69
Orthostatic Intolerance Upon Return to Earth Gravity, 69
Radiation Hazards, 70
Physiological Effects of Stress, 70
Psychological and Social Issues, 70
References, 71

11 PROGRAMMATIC AND POLICY ISSUES 72
International Space Station: Utilization and Facilities, 72
Countermeasure Testing and Validation, 73
  Controlled Trials in Space, 74
  Use of Historical Data, 75
  Empirical Observation, 75
Operational and Research Use of Biomedical Data, 76
  The Role of Medical Operations in Human Research and Countermeasure Validation, 76
  Effects of Spaceflight on Drug Efficacy and Pharmacokinetics, 77
  Availability of Stored Clinical Samples, 77
  Data Archive, 77
Science Policy Issues, 78
  Support of Operational Research, 78
  International Cooperation, 78
  Integration of Research Activities, 79
References, 80

APPENDIXES

A *A Strategy for Research in Space Biology and Medicine in the New Century,*
   Executive Summary 83
B Letter of Request from NASA 100
C Glossary 103
D Acronyms 107
E Biographies of Committee Members 110

Executive Summary

The 1998 Committee on Space Biology and Medicine (CSBM) report *A Strategy for Research in Space Biology and Medicine in the New Century* (NRC, 1998) assessed the known and potential effects of spaceflight on biological systems in general and on human physiology, behavior, and performance in particular, and recommended directions for research sponsored over the next decade by the National Aeronautics and Space Administration (NASA). **The present follow-up report reviews specifically the overall content of the biomedical research programs supported by NASA in order to assess the extent to which current programs are consistent with recommendations of the *Strategy* report for biomedical research activities.** In general, NASA programs concerned with fundamental gravitational biology are not considered here. The committee also notes that this report does not include an evaluation of NASA's response to the *Strategy* report, which had only recently been released at the initiation of this study.

Summarized below are the committee's findings from its review of (1) NASA's biomedical research and (2) programmatic issues described in the *Strategy* report that are relevant to NASA's ability to implement research recommendations.

NASA BIOMEDICAL RESEARCH

Most of the biomedical research funded by NASA is carried out through (1) a program of NASA Research Announcements that funds proposals by individual investigators, (2) research conducted in scientific or clinical programs at either the Johnson Space Center (JSC) or the Ames Research Center (ARC), and (3) focused research projects managed by the National Space Biomedical Research Institute (NSBRI). The committee considered all NASA biomedical research projects, irrespective of their origin, under the following disciplinary categories:

- Sensorimotor integration,
- Bone physiology,

- Muscle physiology,
- Cardiovascular and pulmonary systems,
- Endocrinology and nutrition,
- Immunology and microbiology,
- Radiation biology, and
- Behavior and performance.

In order to assess the degree to which NASA's research programs will meet the agency's needs for biomedical knowledge in the next 10 years, the committee compared current and planned research to the recommendations made in the *Strategy* report. Within this context, the committee attempted to answer the following questions.

What Is the Balance of Discipline Areas in NASA's Biomedical Research Program?

The *Strategy* report gave the highest overall priority to specific research questions dealing with bone and muscle loss, changes in the function of the vestibular and sensorimotor systems, orthostatic intolerance, radiation hazards, and the physiological and psychological effects of stress. Although the committee found the balance of NASA research between the various biomedical disciplines to be generally consistent with the relative emphasis given to them in the *Strategy* report, many of the specific research topics given the highest overall priority are still to be addressed. Noted below is the degree to which these research topics appear in the current program. It should be kept in mind that many of the *Strategy* report recommendations called for specific microgravity investigations that cannot be carried out until appropriate flight opportunities again become available.

As recommended, mechanistic studies and the use of ground-based animal models to understand changes in bone and muscle during and after spaceflight are being emphasized in NASA's current program. Preliminary ground studies of the relationship between exercise activity and protein-energy balance have also been started. Implementation of recommendations to collect in-flight astronaut data on bone loss and hormonal profiles must await flight opportunities.

Some preliminary investigations have been carried out that are relevant to the recommendation for in-flight recordings of signal processing following otolith afferent stimulation. However, the recommendation to study the basis for compensatory vestibulomotor mechanisms on Earth and in space has not yet been addressed. The performance of the recommended microgravity studies on neural space maps and pattern learning in the vestibulo-oculomotor system will depend on the availability of flight opportunities.

Mechanistic studies of total peripheral resistance responses during postflight orthostatic stress have been conducted on the recent Neurolab mission and in the cardiovascular laboratory at JSC. The Mir cardiovascular experiments were relevant to the recommendation to examine cardiovascular changes on long-duration missions. However, inadequate plans exist to monitor these changes on the International Space Station. Current pulmonary studies focus on the issue of decompression sickness but do not address aerosol deposition and respiratory muscle function.

Studies to examine the space radiation-induced risks of cancer and central nervous system damage are being carried out by NSBRI investigators at new facilities at Loma Linda University for proton studies and at Brookhaven for heavy ions. These will provide greatly improved access to investigators for relevant studies. Flights are not yet available for the recommended study of the combined effects of radiation and stress on the immune system, and no preliminary ground studies on this issue appear to be planned.

The majority of NASA-supported psychosocial research is currently directed toward the recommended studies of neurobiological mechanisms involved in circadian rhythm and sleep disturbances, and there are strong indications that NASA also plans to give explicit emphasis to the recommended studies on psychosocial mechanisms in the future. However, the work recommended on countermeasure evaluation and development has so far received little attention, with the exception of circadian and neurovestibular system studies.

What Is the Balance Between Ground and Flight Investigations?

The majority of NASA's current and planned biomedical studies are ground based due to the limitation in flight opportunities over the next several years. Some notable exceptions include behavioral research and sensorimotor integration, which have a significant percentage of experiments in the flight program. As for radiation biology, its program focus on ground-based research is consistent with the recommendations of the *Strategy* report. However, the *Strategy* report recommended a major flight component for most discipline research programs, and the current lack of appropriate flight opportunities may lead to delays in the development of needed countermeasures for physiological changes such as orthostatic intolerance and muscle loss. Although it is possible, and even necessary, to perform much of the preliminary work on the ground, many of the critical research questions cannot be resolved without in-flight studies.

To What Degree Are Studies of Fundamental Cellular and Physiological Mechanisms Addressed in Research Programs?

In general, there is a strong and very appropriate degree of emphasis on mechanistic studies across the various biomedical disciplines, as recommended in the *Strategy* report. In the area of bone physiology, for example, an independent program of basic cellular and molecular biology has been initiated at ARC, while an NSBRI laboratory is taking pharmacologic approaches to the study of biochemical pathways. Some of the specific mechanistic studies recommended in the *Strategy* report remain to be addressed, however, with studies of psychosocial mechanisms being particularly sparse.

What Are the Plans for Validation of Animal Models?

In most of the disciplines for which a need for animal research was cited in the *Strategy* report, NASA is making significant use of animal models. However, their use in sensorimotor integration studies is thus far limited to only a few of the recommended research topics. One widely utilized animal model is hindlimb unloading in rodents, which is being used to study muscle atrophy, bone loss, and immunological changes.

The extent to which the various animal models are being tested to confirm that they duplicate certain physiological changes seen in space-bound humans was more difficult for the committee to determine. However, it is known that attempts have been made, or are planned, to validate aspects of the models used in bone and immunology studies. It was noted that evaluation is needed of the models used in studies of cardiovascular adaptation and endocrine changes.

What Are the Plans for the Development and Validation of Physiological and Psychological Countermeasures?

Although NASA has countermeasures in place for a number of the adverse effects of spaceflight on humans, many have not been rigorously tested for efficacy and side effects. However, the development of future countermeasures is the primary focus of NSBRI research, and JSC is developing an administrative mechanism for soliciting and testing countermeasures. Issues related to implementing this process are discussed under programmatic issues below.

Most of NASA's discipline programs include some level of research activity directed at adverse effects, such as bone and muscle loss, for which no effective countermeasure exists. Studies include investigation of the respective effects of pharmacologic intervention, nutrition, and centrifugation on bone loss, renal stone formation, and sensorimotor impairment. However, there is considerable variation in the organization and scope of these activities. Although studies on respiratory tract infections appear likely to meet countermeasure goals, planning appears to be very limited for developing and testing countermeasures for orthostatic intolerance, psychosocial deficits, and radiation effects.

What Are the Plans to Perform Epidemiology and Monitoring?

Plans exist to monitor indicators for a number of physiological changes in International Space Station (ISS) astronauts. These include measurements of bone density change, radiation exposure, orthostatic intolerance, and cardiac atrophy after missions longer than 30 days. Muscle atrophy will be monitored as part of a program of countermeasure testing, and the capability for in-flight monitoring of psychological status is planned. There are, however, a number of factors that may limit the usefulness of the collected data. Much of the data is collected for medical operations purposes and will not be accessible to the scientific community, nor in many cases do there appear to be plans even within the clinical program to systematically analyze and interpret the data. In addition, it is not clear that in every discipline the techniques best suited to the measurement, such as the use of magnetic resonance imaging (MRI) to measure cardiac mass, will be used on a routine basis.

To What Extent Are Programs Supporting New, Advanced Technologies and Methodologies?

Considerable attention is being paid to the development of new technologies and methodologies that can be used in basic research, monitoring of in-flight physiological changes, and countermeasures. This seems to be true across nearly all discipline programs and in all components of the program. It is a particular focus of work at NSBRI. Some of the more innovative approaches under development include a portable bone densitometer, virtual environments to study human perception and navigation, and advanced telemetric-based sensor systems.

PROGRAMMATIC ISSUES

The 1998 *Strategy* report raised a number of concerns in the program and policy arena, including issues relating to strategic planning, conduct of space-based research, and utilization of the ISS; mechanisms for promoting integrated and interdisciplinary research; and collection of and access to human flight data. The committee looked at both the current program and what was known regarding future plans in order to evaluate the congruence with *Strategy* report recommendations. Additional overarching issues having to do with countermeasure testing and validation, and with the role of the Office of Medical Operations in human research, came to the committee's attention during the course of the present study. Some of the most significant issues that remain to be addressed follow.

International Space Station: Utilization and Facilities

The adequacy of the life sciences research facilities that will actually be in place on the ISS at its final build-out remains an issue of serious concern. Possible design changes, the mounting delays in utilization timetables, and the perceived potential for downgrading of research facilities and budgets have continued to erode the confidence of the user scientific community. Important questions also remain about the availability of Russian cosmonauts for long-term follow-up in the conduct of biomedical research, especially in the early phases of ISS utilization.

Countermeasure Testing and Validation

The need for effective countermeasures against the deleterious effects of spaceflight on astronaut health and performance will become increasingly critical as longer-duration flights become the norm on the ISS and beyond. The development of effective, mechanism-based countermeasures requires three well-integrated phases: (1) basic research to identify and understand mechanisms of spaceflight effects; (2) testing and evaluation of proposed countermeasures to determine their efficacy; and (3) validation of promising countermeasures by well-designed clinical studies. Recently, NASA has begun to develop a standard procedure for testing and evaluating countermeasures, but this has not yet been implemented. It is essential that the process, once in place, be readily accessible to all investigators, extramural as well as intramural, and that criteria for acceptance into the testing program be clearly defined.

Operational and Research Use of Biomedical Data

Access to in-flight biomedical data, as well as to longitudinal data collected during postflight longitudinal monitoring of astronaut health, is limited, and the partial and incomplete availability of human data to qualified investigators was highlighted as a major concern in the *Strategy* report and continues to be an issue. The committee urges that NASA explore ways in which these data and samples, collected in the past and future, can be made available to investigators. Additionally, steps are needed to ensure that future data collection includes measurements and sampling that have been optimized to give the most useful information on in-flight development of problems and postflight recovery of normal physiological function. The role played by the crew surgeon is especially critical to collection of these data, and rigorous training in clinical research and basic research is recommended as a requirement for the position.

POLICY ISSUES

International Cooperation

The era of ISS construction and utilization, with increased emphasis on international crews and operations, raises important issues with respect to acquisition and management of human data. Mechanisms are needed to ensure that protocols and facilities for pre- and postflight monitoring and testing are consistent across national boundaries. There must be common criteria for evaluation and utilization of countermeasures and international cooperation in their development.

Integration of Research Activities

NASA funding for biomedical research is increasingly distributed among a diverse set of organizations and programs. These include the program of NASA Research Announcements (NRAs), intramural investigators in NASA center science programs, the NSBRI, and the NASA Specialized Centers of Research and Training. NASA science benefits from the unique strengths of each of these program constituents, but careful planning is required to delineate the roles, responsibilities, and appropriate funding levels for each; to ensure effective collaborations; and to integrate research findings. In particular, NASA should maintain a healthy NRA program as the primary mode for support of space-related biomedical research because it remains the best method of accessing the entire investigator community and exploring novel ideas and approaches.

REFERENCE

National Research Council (NRC), Space Studies Board. 1998. A Strategy for Research in Space Biology and Medicine in the New Century. Washington, D.C.: National Academy Press.

1

Introduction

The 1998 Committee on Space Biology and Medicine (CSBM) report *A Strategy for Research in Space Biology and Medicine in the New Century* (NRC, 1998) assessed the known and potential effects of spaceflight on biological systems in general and on human physiology, behavior, and performance in particular, and recommended directions for research sponsored by the National Aeronautics and Space Administration (NASA) over the next decade. The present follow-up report reviews specifically the overall content of the biomedical research programs supported by NASA in order to assess the extent to which current programs are congruent with recommendations of the *Strategy* report for biomedical research activities.

NASA biomedical science includes the NASA Research Announcement (NRA) program of NASA headquarters' Division of Life Sciences, the National Space Biomedical Research Institute (NSBRI), and the Ames Research Center (ARC) and Johnson Space Center (JSC). NRAs are made annually and include guidelines regarding priority areas of research for the upcoming year. The NRA program is open to all qualified investigators, who submit investigator-initiated proposals; funding is based on peer review and considerations of program relevance. NSBRI, a recently established consortium of university- and NASA center-based investigators, receives core support from NASA; participating investigators also compete for NASA funding through the NRA program, as well as for support from the National Institutes of Health (NIH) and other extramural funding agencies. At present, NSBRI focuses on ground-based research, with an emphasis on basic and applied research relevant to the development of improved countermeasures. Biomedical research at ARC emphasizes neurovestibular studies and psychosocial issues. In addition, ARC houses unique centrifuge facilities for hypergravity studies in humans and animals. Biomedical activities at JSC are included in the Astronaut Office and Space Life Sciences Directorate, and ground-based testing of proposed countermeasures is the responsibility of this center. NASA also supports a number of focused research programs at universities, called NASA Specialized Centers of Research and Training (NSCORT). Currently one biomedical NSCORT, focused on radiation health research, is being funded.

The NRA program and NSBRI core funding provide the major sources of support for space biomedical research, although human studies are a significant component of the Space Life Sciences Directorate at JSC. The headquarters NRA program supports investigator-initiated ground-based and spaceflight research in all relevant disciplines through a universal peer review process to which NASA intramural scientists as well as extramural investigators can apply. Disciplinary research carried out at both Ames and JSC is funded by this program and is summarized in the relevant disciplinary chapters of this report. The primary mission of NSBRI-funded investigators is basic research aimed at the development of countermeasures, while countermeasure evaluation and testing are assigned to and supported by JSC. Program oversight is carried out by the headquarters Office of Life Sciences, although NSBRI operates with a high degree of autonomy.

For the present report, the committee was charged to map the current NASA-supported biomedical research program to the recommendations of the *Strategy* report, not to evaluate in detail the quality of the research being conducted. Thus, this report presents findings and conclusions but makes few specific recommendations beyond those contained in the 1998 *Strategy* report. These findings and conclusions are highlighted in the body of this report. It should be emphasized that the *Strategy* report was released only in September 1998, and the "current" program, as detailed in 1998 and 1999 program documents, was therefore put in place prior to issuance of that report. The FY 2000 NRA and NSBRI research solicitations represent the first opportunities for NASA to respond to specific recommendations of the *Strategy* report, and the programmatic priorities indicated in these announcements are discussed in discipline-specific chapters of the present report.

The 1998 *Strategy* report assessed research needs in the broad spectrum of disciplinary areas relevant to the health and performance of astronauts in space, ranging from cell and developmental biology, through the major physiological systems—bone and muscle, cardiopulmonary, endocrinology and nutrition, and immunology and microbiology—to radiation hazards and issues related to behavior and performance, and made specific disciplinary recommendations in each. In addition, the report considered overall biomedical research priorities and recommended that the highest priority for NASA-supported biomedical research be given to problems that may limit astronauts' ability to survive and/or function in prolonged spaceflight. Six issues were identified: (a) loss of weight-bearing bone and muscle; (b) vestibular function, the vestibulo-ocular reflex and sensorimotor integration; (c) orthostatic intolerance upon return to Earth gravity; (d) radiation hazards; (e) physiological effects of stress; and (f) psychological and social issues. A wide range of approaches was recommended, aimed at understanding fundamental mechanisms of effects induced by spaceflight and the development of effective, mechanism-based countermeasures.

In order to assess the extent to which the existing overall biomedical research program is aligned with the recommendations of the *Strategy* report, the committee considered the number of specific projects and the range of research activities that were ongoing within the disciplinary and subdisciplinary components of the NRA program and NSBRI for FY 1998 and FY 1999 funding years. In addition, the current total discipline and subdiscipline funding was estimated as a rough index of the priority given to the various research areas. Differences in grant accounting methods as well as overlaps in disciplinary research content among programs made it difficult in some cases for the committee to define precise budgetary figures. The total NASA funding in FY 1999 for the conduct of research in the Biomedical Research and Countermeasures (BR&C) program, which, in addition to biomedical NRA grants and core support for NSBRI, includes funds for the radiation NSCORT and biomedical support projects, was approximately $36 million. A pie chart summarizing NASA's estimate of budget figures for disciplinary programs within BR&C in FY 1999 is presented in Figure 1.1. Total FY 1999 NSBRI research funding, most of which was provided by BR&C, was approximately $1 million each for the eight areas

INTRODUCTION

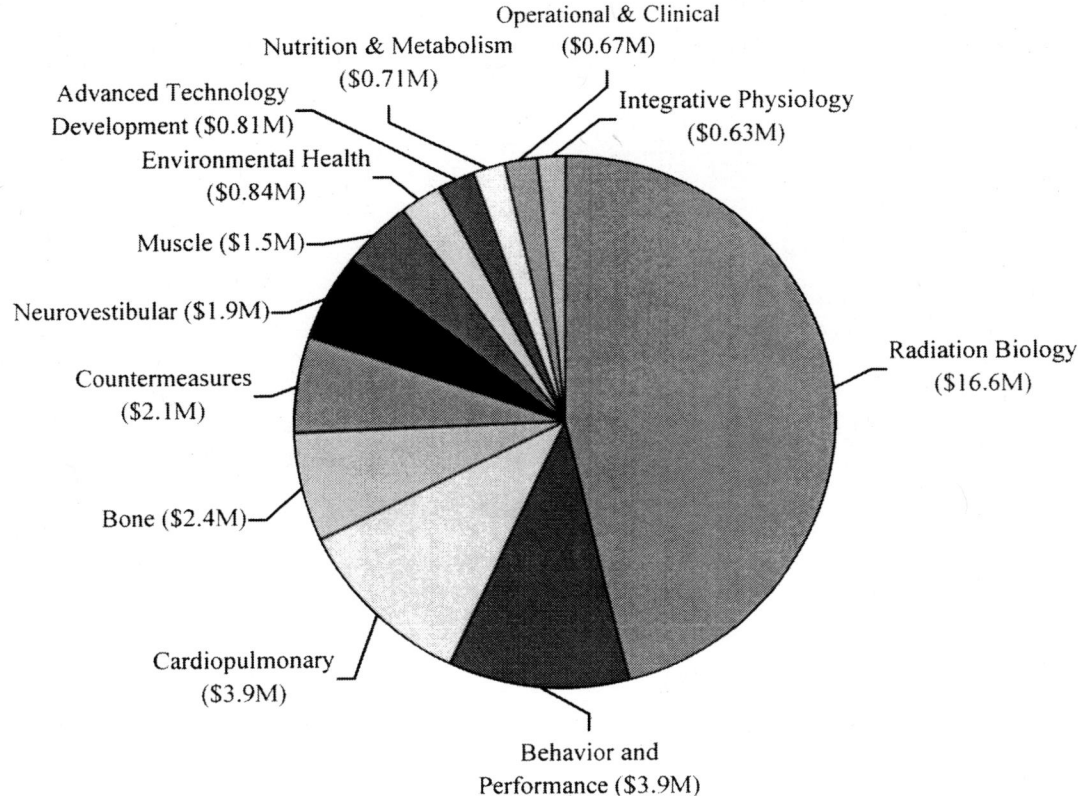

FIGURE 1.1 FY 1999 funding of NASA-supported programs in biomedical research and countermeasures. The total FY 1999 budget for biomedical research and countermeasures was approximately $36 million. It should be noted that the radiation biology sector includes major facility construction costs and directed spending not incurred in other disciplines. The actual budget for peer-reviewed research in radiation health was roughly $4.6 million. Details of the radiation biology budget appear in Table 8.1.

of research focus—bone, cardiovascular, human factors, immunology-microbiology-hematology, muscle, neurovestibular, radiation, and technology development. In certain areas of research, additional limited amounts of funding may be derived from sources not included here, such as NASA's Fundamental Biology Program or joint projects with other agencies. Where information was available on these projects it is noted in the detailed discussions of individual disciplines in the chapters that follow. The Fundamental Biology Program[1] per se lay outside the charge to the committee; however, projects relevant to biomedical science that are funded under this program are included in the review of the current biomedical research program overall.

This report is organized similarly to the *Strategy* report. The body of this report addresses the major physiological systems, radiation effects, and biobehavioral aspects known or likely to be affected by spaceflight and describes the following elements for each:

---

[1] Formerly known as the Gravitational Biology and Ecology program.

- NASA's current research program and the extent of its congruence with recommendations of the *Strategy* report;
- Programmatic balance:
  —The balance between subdiscipline areas emphasized in the current program,
  —The balance between ground and flight investigations, and
  —The emphasis given to fundamental mechanisms;
- Utilization and validation of animal models;
- Development and validation of countermeasures;
- Epidemiology and monitoring; and
- Support of advanced technologies.

The report then describes the extent to which the current program addresses the recommendations of the *Strategy* report for the highest-priority research to address critical questions relating to the physiological and psychological effects of spaceflight. A final chapter discusses continuing programmatic and policy concerns relating to strategic planning, conduct of space-based biomedical research and utilization of the International Space Station, mechanisms for promoting integrated and interdisciplinary research, and the collection and utilization of human flight data for research purposes. In addition, the committee discusses issues having to do with countermeasure testing and validation and the role of the Space Life Sciences Directorate in human research that came to the committee's attention during the course of the present study.

REFERENCE

National Research Council (NRC), Space Studies Board. 1998. A Strategy for Research in Space Biology and Medicine in the New Century. Washington, D.C.: National Academy Press.

2

Sensorimotor Integration

INTRODUCTION

Sensorimotor integration plays a critical role in the control of posture and movements, and so is essential for locomotion and using tools. On exposure to an altered gravitational environment, there are changes in the sensory signals originating from the vestibular system, particularly those coming from the otolith organs. These changes have major effects on visual and spatial orientation, and on mobility during spaceflight and on return to Earth. Disturbances occur more frequently on longer-duration missions, and complete recovery can take weeks and, in some cases, months. Disorientation, impaired visual acuity, and postural instability can have profound effects on the performance of sensorimotor tasks, including piloting the spacecraft or making an emergency egress. Accordingly, ground-based research studies funded by the National Aeronautics and Space Administration (NASA) have focused on examining the time course of adaptation of the gravity-sensitive properties of the vestibulo-ocular, vestibulo-collic, and vestibulo-spinal reflexes and on determining how vestibular, proprioceptive, and visual information is used to control head and trunk position, maintain postural stability, and assist in locomotion. In addition to studying these activities in normal subjects, research is also performed on subjects with pathological changes in the neurovestibular system, as symptoms in these latter cases often mimic the changes observed in astronauts during and after spaceflight.

Early in-flight performance for many astronauts is often marked by space motion sickness, particularly for astronauts with no previous spaceflight experience. In long-duration flights, space motion sickness can also be triggered by illusions of visual reorientation that occur when astronauts misperceive their orientation with respect to the environment.

The chapter "Sensorimotor Integration" in the *Strategy* report (NRC, 1998) defines six priority areas of investigation: (a) spatial orientation; (b) posture and locomotion; (c) vestibulo-ocular reflexes (VORs) and oculomotor control; (d) space motion sickness; (e) central nervous system (CNS) reorganization and vestibular processing during microgravity; and (f) teleoperation. In addition, the chapter "Developmental Biology" in that report contained three sections pertinent to sensorimotor integration: (a) development of the vestibular system; (b) neural space maps; and (c) neuroplasticity. Grants focused on these last three topics are funded primarily through the Gravitational Biology and Ecology (GB&E)[1] program and have been included in this review. The current NASA biomedical research program is considered in light of all nine subdiscipline areas. Research projects that have a primary focus on the interactions of the vestibular system with the cardiovascular or musculoskeletal systems are not covered here.

NASA'S CURRENT RESEARCH PROGRAM IN SENSORIMOTOR INTEGRATION

Altogether in FY 1999, NASA funded 39 projects relevant to sensorimotor integration; these are classified in Table 2.1 according to the nine subdiscipline areas described above. Funding for this research was contributed by three NASA programs (Biomedical Research and Countermeasures, Advanced Human Support Technologies, and Gravitational Biology and Ecology), the National Space Biomedical Research Institute (NSBRI), and a new cooperative National Institutes of Health (NIH)-NSBRI program. Within the Biomedical Research and Countermeasures (BR&C) program, sensorimotor integration research is included in three disciplines: (1) behavior and performance; (2) physiology: neuroscience; and (3) operational and clinical. The 12 projects funded through the GB&E program are concerned with the basic mechanisms underlying vestibular function or the effects of gravity on vestibular development. In FY 1999, the total funding for research on sensorimotor integration was approximately $7.8 million, which places this work in the top quarter of NASA's Biomedical Research (NASA, 1999). In FY 1998, the funding level for sensorimotor integration research was approximately $6 million, with 40 projects being supported. The increase in funding in FY 1999 reflected the new cooperative NIH-NSBRI program for the support of vestibular research, which had a funding level of $1.4 million, 80 percent of which was provided by the National Institute of Deafness and Communicative Disorders.

Studies on sensorimotor integration are carried out in all four components of the NASA life sciences research enterprise, including the laboratories at Johnson Space Center (JSC), Ames Research Center (ARC), and numerous universities funded either through the NASA Research Announcement (NRA) mechanism or through the NSBRI. In addition to conducting intramural research, scientists at JSC and ARC routinely collaborate on university-based research programs. JSC scientists have also been involved in setting up laboratories in Star City, Russia, to investigate postural control, locomotion, and visual target acquisition after long-duration spaceflight. The Neurological Function Section at JSC is focused on human studies of posture and locomotion, vestibulo-ocular reflex and oculomotor control, and visual adaptation and space motion sickness. Specialized facilities exist for pre- and postflight studies of visuospatial adaptation and postural stability. At ARC, both human and animal studies are conducted on spatial orientation, vestibulo-ocular reflexes and oculomotor control, and space motion sickness. ARC has facilities for centrifuge studies on humans to determine the time course of adaptation to different g-levels and the effects of altered gravity on human behavior and performance. The center also has several facilities for testing the responses of the vestibular system to linear and angular acceleration in small animals and in human subjects.

There are three major research projects supported entirely by the NSBRI in the area of neurovestibular adaptation, two of which have a coinvestigator at JSC. These are focused on adaptation of vestibular reflexes to different gravitoinertial force conditions; spatial orientation and mobility, in particular visually induced tilt and reorientation illusions; and eye, head, and body movements during locomotion and their stability in a range of environments. A fourth project on neurovestibular adaptation

---

[1] Recently retitled the Fundamental Biology Research Program (FBRP).

TABLE 2.1 Summary of Funding in FY 1999 for Sensorimotor Integration Subdisciplines

Subdiscipline	NRA Total ($ thousands)	NRA No. of Projects	NSBRI Total ($ thousands)	NSBRI No. of Projects	NSBRI-NIH Total ($ thousands)	NSBRI-NIH No. of Projects
Spatial orientation	1,188	5	339	1	112	1
Posture and locomotion	1,024	5	369	1	0	0
VOR and oculomotor control	962	6	321	1	1,301	5
Space motion sickness	130	1	0	0	0	0
CNS reorganization and vestibular processing during microgravity	555	3	0	0	0	0
Teleoperation	179	1	0	0	0	0
Development of vestibular system	706	6	0	0	0	0
Neural space maps	84	1	0	0	0	0
Neuroplasticity	514	2	0	0	0	0
Total	5,342	30	1,029	3	1,413	6

supported by the NSBRI is an interdisciplinary proposal (i.e., identified as a synergy project) concerned with the visual and vestibular autonomic influence on short-term cardiovascular regulatory mechanisms.

In the BR&C program, research on sensorimotor integration is heavily concentrated on human studies of postural stability, the vestibulo-ocular reflex and oculomotor control, and visuospatial orientation and adaptation. One-third (6/18) of the projects in the BR&C program are focused on postural control and motor adaptation to variations in gravitational force level, one of the issues identified in the *Strategy* report. The other two recommendations made in the *Strategy* report for future studies of posture and locomotion concern the development of ancillary sensory aids to facilitate postural and locomotory control and to assist in adaptation during transitions between gravitational force environments, and the development of animal models of reentry disturbances. These do not appear to be the primary focus of any studies being funded by NASA at present.

Studies of oculomotor control and vestibular reflexes account for approximately one-third of the sensorimotor integration projects funded through NASA's BR&C program and represent a major emphasis of the program. These projects address all of the issues raised with respect to this area in the *Strategy* report. This research has contributed to the understanding of the effects of microgravity on the vestibulo-ocular system, on the control of head and eye position, and on how the vestibular reflexes adapt to altered gravitational conditions.

Accurate spatial orientation is essential in order to control one's movements and to interact with objects in the environment. On Earth, gravity plays a fundamental role in spatial orientation. In the absence of gravitational cues during spaceflight, astronauts often misinterpret visual information and experience visual reorientation illusions when they misperceive their own orientation with respect to the environment. Six projects in the BR&C program are concerned with spatial orientation, which is often studied in conjunction with gaze stability or postural and locomotor control. This research includes both flight- and ground-based studies and is consistent with most of the recommendations made for research on spatial orientation in the *Strategy* report. In particular, there are human studies on the identification of sensory, motor, and cognitive factors that influence the ability to adapt to different gravitational environments, as well as animal research on the neural coding of spatial mobility. One recommendation concerning the influence of microgravity on the integrative coordination of active movements before, during, and after spaceflight does not appear to be the subject of study at this time.

Space motion sickness can be an operational problem during the first 72 hours of flight and can be controlled with intramuscular injections of promethazine. It can also be triggered during spaceflight by visual reorientation illusions, which occur when crew members misinterpret visual cues from surrounding objects, usually when they are working in an unfamiliar "agravic" orientation. Only one project specifically devoted to space motion sickness is funded through NASA's BR&C program, and it is concerned with the effects of promethazine on human performance. There are a few structural and functional studies on the interactions between the vestibular system and autonomic function, including cardiovascular regulation, which address one of the four recommendations made for space motion sickness in the *Strategy* report. Another project examines the possibility of maintaining dual adaptations to more than one force background, a subject recommended for study in the *Strategy* report. Other recommendations made in the *Strategy* report, including studies of the relation between motion sickness and altered sensorimotor control of the head and body; the time course of the Sopite syndrome (i.e., a form of motion sickness associated with prolonged exposure to unusual gravity conditions, whose primary features include drowsiness, fatigue, lack of initiative, apathy, and irritability); and the relation between terrestrial and space motion sickness have not yet been investigated.

Studies on vestibular signal processing and development funded through the GB&E program (n = 12) represent approximately one-third of the total number of research projects included in the topic of sensorimotor integration. The research is focused on measuring the changes in signal processing by peripheral vestibular afferents in altered gravitational environments and on the effects of gravity on vestibular development using a wide range of animal models (see discussion of validation of animal models below). The research is focused primarily on structure-function studies of the peripheral vestibular system. Findings on the central vestibular system are lacking so far, although they are cited in a presently funded proposal. With respect to the recommendations made in the *Strategy* report, as yet there are no studies focused on the development of the central vestibular system or on how exposure to microgravity affects neuroplasticity and the different space maps located in the brainstem, sensory and motor cortices, and corpus striatum.

The *Strategy* report made two recommendations concerning studies of vestibular processing: (1) study the effect of altered calcium regulation in microgravity on otoconial development and regeneration in animal models and (2) perform in-flight electrophysiological recordings from otolith afferents and efferents, and investigate signal processing within the central vestibular pathways in animals. During the recent Neurolab flight, some studies were conducted on otoconial development in microgravity and preliminary electrophysiological recordings of otolith afferents were performed in fish.

Consistent with the recommendations made in the *Strategy* report, research in the Biomedical Research and Countermeasures program is focused heavily on human studies, both flight and ground based, of postural stability, the vestibulo-ocular reflex and oculomotor control, and visuospatial orientation and adaptation. Studies of vestibular signal processing and development constitute about one-third of the total number of research projects included in sensorimotor integration research and are performed on a wide range of animal models. In keeping with the

recommendations of the *Strategy* report, this work is concerned with investigating signal processing by peripheral vestibular afferents in altered gravitational environments and the effects of gravity on vestibular development. However, other recommendations regarding the development of the central vestibular system and how exposure to microgravity affects neuroplasticity and neural space maps are not being adequately addressed at present.

PROGRAMMATIC BALANCE

Balance of Subdiscipline Areas

NASA's program of research on sensorimotor integration places a major and appropriate emphasis on characterizing the vestibular-related disturbances in visual and spatial orientation and posture observed in astronauts during and after spaceflight. Much of this work is performed on normal human subjects. Structure-function studies of the peripheral vestibular system represent a small but active component of the research supported by NASA on sensorimotor integration. Other areas, such as CNS reorganization and teleoperation, have not received adequate attention to date.

Balance of Ground and Flight Investigations

The total number of sensorimotor integration-related projects (39) is almost evenly divided between ground and flight research, which is appropriate. In FY 1999, the Advanced Human Support Technologies program (n = 3) funded only flight-based projects, whereas the Biomedical Research and Countermeasures program (n = 18) supported 7 flight and 11 ground-based studies and the Gravitational Biology and Ecology Program (n = 12) supported 5 ground-based and 7 flight projects (NASA headquarters). These 7 flight projects were primarily a continuation of Neurolab experiments or the analyses of data collected during that mission. Despite the paucity of flight opportunities due to the construction of the ISS, there are several funded projects that will be flight-borne when feasible.

Emphasis Given to Fundamental Mechanisms

There is an active research program directed toward elucidating the fundamental mechanisms underlying the function of the vestibular system. As yet, the human and animal studies are not well integrated, and findings from human research do not lead naturally to further exploration of fundamental mechanisms in experimental animal models or vice versa. Specifically, the findings from studies on basic mechanisms are not applied to derive new countermeasures that could improve astronauts' performance and/or safeguard their health.

Utilization and Validation of Animal Models

Excluding the Neurolab projects, animal research receives less than 25 percent of the funds allocated for studies on sensorimotor integration and thus constitutes a relatively small but significant part of NASA's biomedical research program. The research projects involving experimental animals are predominantly ground based and are performed mainly in university laboratories. This research is focused on the effects of gravity on signal processing and adaptation in the peripheral vestibular system and vestibular development using a wide spectrum of species including primates, rodents, birds, amphibians, fish, and mollusks. There are no animal studies on vestibular compensation in microgravity.

Although studies of vestibulo-oculomotor learning are being carried out on human subjects, no animal models have been investigated.

NASA's program of research on sensorimotor integration places a major and appropriate emphasis on characterizing the vestibular–related disturbances in visual and spatial orientation and posture observed in astronauts during and after spaceflight. These projects are almost evenly divided between ground and flight research, which is reasonable. Other areas identified in the *Strategy* report such as CNS reorganization and teleoperation have not received adequate attention to date. Although sensorimotor integration studies are focused heavily on human subjects, there is an active animal research program directed toward elucidating the fundamental mechanisms underlying the function and dysfunction of the vestibular system. As yet, human and animal studies are not well integrated with regard to understanding basic mechanisms underlying vestibular adaptation.

DEVELOPMENT AND VALIDATION OF COUNTERMEASURES

Countermeasures considered effective for some of the sensorimotor alterations that occur due to spaceflight include the administration of promethazine, crew training, time-line adjustments (i.e., not scheduling certain activities such as extravehicular activities (EVAs) during the first few days of spaceflight), timing of training with respect to the mission start, in-flight exercise, and assisted egress (NASA, 1997). Promethazine, an antimotion sickness drug, has been the treatment of choice in the Space Shuttle program for many years, although it does induce adverse side effects on human performance when tested on the ground (Harm et al., 1999). As yet, there are no systematic studies of the effects of promethazine on human performance in space, although one project has been funded. Most of the data regarding the lack of sedative side effects in space are anecdotal. Other pharmacologic treatments for motion sickness (e.g., selective muscarinic antagonists) warrant investigation as prophylactic treatment for space motion sickness.

Clearly, training and exposure to altered gravitational environments, either on Earth or during previous missions, assist astronauts in adapting to spaceflight. These factors are cited as effective countermeasures (NASA, 1997). At present, it is unclear what specific aspects of this training facilitate adaptation or compensation. An essential prerequisite for developing virtual reality simulators is to identify which elements of the astronaut's training facilitate adaptation to an altered gravitational environment and maintenance of spatial orientation in a spacecraft. There is evidence that previous spaceflight experience or repeated exposure to altered gravitational environments reduces the frequency of space motion sickness, spatial orientation problems, and disequilibrium. However, it is unknown whether this is accomplished in increments or whether steady state is achieved after a certain number of exposures.

Centripetal acceleration induced by centrifugation has been proposed as an in-flight sensorimotor countermeasure to promote appropriate responses to sustained changes in gravitoinertial forces. Several research studies have been conducted with human subjects at JSC, ARC, and university laboratories to study the mechanisms of adaptation to altered gravitational environments and to examine postural stability following centrifugation. The ongoing exploration of artificial gravity as a possible countermeasure for sensorimotor impairment merits continued attention. However, the benefits derived by other systems (e.g., musculoskeletal) from exposure to artificial gravity may be limited ultimately by the adaptive capabilities of the vestibular system. It is not clear whether humans can maintain dual adaptations to different gravitational environments, although this is being studied at present.

Human mobility in the spacecraft has not been studied in microgravity. Anecdotal reports suggest that astronauts' experiences in mock-ups, parabolic flights, neutral buoyancy, and virtual reality simulators assist them in orientation and mobility, but these have not been studied systematically to determine their effectiveness as countermeasures. Clearly, considerable individual differences exist in the ability to orient and navigate and to remain oriented using visual cues. There exists a substantial body of operational and experimental data on orientation, mobility, and postural control at JSC that has not been analyzed fully. These data could provide answers to some of the questions raised here. The individual variability in response to spaceflight makes it imperative to use as large a sample size as possible and a longitudinal design.

Countermeasures that have been implemented to deal with the sensorimotor alterations that occur due to spaceflight include providing anti-motion sickness drugs and modifying the duration and timing of crew training. There is evidence that previous spaceflight experience or repeated exposure to altered gravitational environments reduces the frequency of space motion sickness, spatial orientation problems, and disequilibrium. However, it is unclear what specific pharmacological countermeasures or aspects of training promote adaptation to or compensation for altered gravitational environments.

EPIDEMIOLOGY AND MONITORING

At present, much of the information available on the frequency and severity of problems affecting the visual-vestibular and postural control systems during and after spaceflight is anecdotal or not accessible. However, data on the in-flight use of medications and on the incidence of space motion sickness have been collected as part of the Longitudinal Study of Astronaut Health. Data on the use of anti-motion sickness medications will continue to be collected since the International Space Station (ISS) Medical Operations Requirements Document (MORD) requires the operational monitoring of countermeasures, such as pharmacological preparations. In addition, the Integrated Testing Regimen (ITR), a standardized set of physiological and psychological tests that will be conducted before, during, and after spaceflight to examine countermeasure efficacy (NASA, 1999), includes a locomotor control test, a gaze holding test, and a functional neurological assessment. The Neurological Function Section at JSC is focused on human studies of sensorimotor integration and has established specialized facilities for pre- and postflight testing of visuospatial adaptation and postural stability. However, it lacks a formal program to measure the performance of astronauts over a prolonged period of time after flight using a variety of visual, spatial, and postural tasks. This type of study would provide information on how rapidly the visual-vestibular and postural control systems readjust to the gravitational environment on Earth and to what extent, and under what conditions, this readjustment may be incomplete.

Much of the available information on the frequency and severity of problems affecting the visual-vestibular and postural control systems during and after spaceflight is anecdotal or not accessible. It is expected that the Integrated Testing Regimen will include a formal evaluation of the visual and postural control systems that will provide information on the efficacy of countermeasures.

SUPPORT OF ADVANCED TECHNOLOGIES

The final two priority areas of study identified in the *Strategy* report are central nervous system reorganization and teleoperation and telepresence. That report identified a need for preliminary studies using functional magnetic resonance imaging (fMRI) pre- and postflight to determine the effects of microgravity on sensory and motor cortical maps, and recommended that strategies be developed to determine the sensorimotor and cognitive consequences of CNS reorganization resulting from exposure to microgravity. These are not currently being done. With regard to teleoperation and telepresence, virtual environments are being developed for use in training and as experimental tools. As these techniques reach the required level of sophistication, their usefulness in controlling equipment and robots remotely will be evaluated. Virtual displays are being developed as part of two projects funded through NASA's Advanced Human Support Technologies program, one of which is based at ARC. Head-mounted see-through displays as well as three-dimensional spatial displays are being developed with the objective of using these to study human perception and mobility. Ultimately, these displays will be used as tools to train astronauts for extravehicular and intravehicular activity, both preflight and also in flight on the ISS using the Virtual Environment Generator.

The *Strategy* report recommended that preliminary studies be conducted on the effects of microgravity on human sensory and motor cortical maps using fMRI and that strategies be developed to determine the sensorimotor and cognitive consequences of CNS reorganization resulting from exposure to microgravity. At present, these are not being done. Virtual environments and head-mounted displays are being developed, and these will be used in teleoperation and for training astronauts.

SUMMARY

NASA's research program has made major advances in characterizing changes in posture and balance experienced by astronauts during spaceflight and in understanding the responses of the visual-vestibular system to altered gravitational conditions. However, as recommended in the *Strategy* report, studies are needed to determine the effects of microgravity on human sensory and motor cortical maps using fMRI, and how microgravity may affect CNS reorganization in general and sensorimotor and cognitive functions in particular. At present, animal studies are supported by NASA to characterize certain basic functions of the peripheral vestibular system in university-based laboratories. Little or no progress has been made on understanding how the central vestibular system is affected by microgravity. Further, there appears to be little, if any, linkage between animal-based and human-based research. In the future, it would be desirable to find a more effective mechanism for integrating information obtained from animal and human studies.

BIBLIOGRAPHY

Harm, D.L., L. Putcha, B.K. Sekula, and K.L. Berens. 1999. Effects of promethazine on performance during simulated shuttle landings. Pp. 148-149 in Proceedings of First Biennial Biomedical Investigators' Workshop, January 11-13, 1999, League City, Texas. Houston, Tex.: NASA and Universities Space Research Association (USRA).

National Aeronautics and Space Administration (NASA). 1997. Task Force Report on Countermeasures: Final Report. Washington, D.C.: NASA.

NASA. 1998. Life Sciences Program Tasks and Bibliography for FY 1998. Washington, D.C.: NASA.

NASA. 1999. Countermeasure Evaluation and Validation Project Plan. Houston, Tex.: Johnson Space Center.

National Research Council (NRC), Space Studies Board. 1998. A Strategy for Research in Space Biology and Medicine in the New Century. Washington, D.C.: National Academy Press.

3

Bone Physiology

INTRODUCTION

Loss of bone constitutes one of the best-documented and oft-studied pathological consequences of exposure to microgravity. Despite more than a decade of research aimed at understanding this phenomenon, elucidation of its cellular and molecular basis remains incomplete, and effective preventive interventions remain elusive. In general terms, bone loss must reflect a disruption of the usual tightly coupled process of bone remodeling. Multiple lines of evidence suggest that exposure to microgravity may lead both to an increased activation of osteoclast-mediated bone resorption and to an inhibition of either the proliferation or the activity of bone-forming osteoblasts. The participation of various stress- and inflammatory-related cytokines, as well as alterations in the normal endocrine regulation of bone cell function, is highly likely. Thus, one hope for effective countermeasures may lie in the use of antiresorptive medications, and another might involve interventions aimed at stimulating bone formation activity. However, whereas several potent inhibitors of bone resorption are approved for clinical use, no approved skeletally anabolic medication is available, leaving mechanical interventions as the only means at present for this purpose.

The *Strategy* report (NRC, 1998) called for a series of both animal and human studies related to microgravity effects on bone loss and recommended several areas of particular emphasis. These included (a) a comprehensive description of the phenomenon in humans, including the development of a careful record of skeletal changes occurring during microgravity and postflight and the time course and rate of these changes; (b) a determination of whether changes produced by microgravity in animal bones mimic human changes and whether they have a similar mechanistic basis, and a comparison of pertinent animal models in spaceflight to ground-based models (such as hindlimb unloading); (c) an investigation of the mechanism of bone loss, including identification of the responsive cells and evaluation of the extent to which bone loss is secondary to systemic effects; and (d) the development of countermeasures for spaceflight and human pathology.

NASA'S CURRENT RESEARCH PROGRAM IN BONE PHYSIOLOGY

In the National Aeronautics and Space Administration's (NASA's) biomedical and countermeasure research budget in FY 1999, about $1.7 million was allocated for grants in the NRA program for studies of bone physiology. An additional $959,000 for bone research was funded through the National Space Biomedical Research Institute (NSBRI) (see Table 3.1). More than a dozen projects in the bone program are being conducted or have been completed recently in laboratories at the Ames Research Center (ARC) and Johnson Space Center (JSC), in four university-based laboratories funded through NSBRI, and in several other university laboratories through extramural grants. Studies range in scope from fundamental cellular biology of bone cells, to biomechanics, to the use of modern technology to assess bone status, to intervention research that could be incorporated into countermeasure programs.

Studies at NASA centers and within NSBRI are aimed at understanding the skeletal response to mechanical loading and its deprivation and the development of countermeasures to ameliorate space-flight-related bone loss. The more basic of these projects are carried out at ARC and NSBRI, whereas interventional research is done at ARC, JSC, and NSBRI. A few service functions of the bone program related to NASA missions are handled exclusively at JSC.

Basic Research

Several basic research initiatives are under way. A system has been developed to study the effects of well-standardized, quantifiable loads applied to cultured bone cells. A program has been initiated to evaluate the histomorphometric and other characteristics of bone loss in hindlimb-unloaded rats. In addition, the assessment of a clinical trial of a combined intervention with exercise loading and a potent bisphosphonate should provide complementarity to ongoing bed rest studies in humans.

Basic programs located at ARC constitute a major component of the overall bone program that should provide valuable information bearing directly on problems identified in the *Strategy* report as being of high priority. These programs include studies of musculoskeletal biomechanics as well as of fundamental bone cell biology. Current projects in bone biology involve the use of contemporary molecular biological techniques to assess the effects of mechanical loads on osteoblastic cells of varying maturity and phenotype. Recently initiated studies involve the application of mechanical loads with complete control and characterization of cycle number, rate, peak intensity, and rate of strain, in order to document changes in osteoblast regulatory genes and gene products.

In accord with *Strategy* report recommendations, one of the NSBRI programs explores the effects of a variety of pharmacologic agonists known to interact with the estradiol, vitamin D, or calcium-sensing receptors on mature bone cells and their precursors. Another addresses the effects of unloading on bone

TABLE 3.1 Summary of FY 1999 Funding for Bone Physiology

Subdiscipline	NRA		NSBRI	
	Total ($ thousands)	No. of Projects	Total ($ thousands)	No. of Projects
Molecular and cellular	625	6	675	3
Countermeasures	1,102	6	284	1
Total	1,727	12[a]	959	4

[a]This number may include projects that were still active but were no longer receiving funding in 1999.

blood flow, incorporating pharmacologic agents to define the possible roles of adrenergic and nitric oxide systems in the regulation of bone circulation. A third program concerns the development and application of newly developed computer algorithms to predict bone strength and fracture risk from noninvasive densitometric data. These algorithms can be applied to both astronaut and animal studies.

Animal Studies

Several animal models for microgravity-associated bone loss have been proposed, with particular attention given in recent years to a rat model involving hindlimb unloading. Although many insights have been obtained using this and other antiorthostatic models of skeletal unloading, concerns remain about their ability to provide a true reflection of what may be experienced by the human skeleton during spaceflight. In particular, it must be recognized that the number and variety of hormonal, nutrient, and other stresses applicable to the human situation may differ fundamentally from those of the hindlimb-unloaded rat. In accord with *Strategy* report recommendations, work at ARC continues to validate the hindlimb-unloading model and to delineate the mechanistic role of various hormones and cytokines. This work should continue to provide valuable information, particularly if molecularly based studies can be incorporated. Also in accord with *Strategy* report recommendations, some studies are exploring the effects of unloading on bone content of bone-specific proteins, and others are developing densitometric methods that should allow accurate descriptions of the nature and distribution of skeletal changes during unloading and spaceflight. New animal work funded in FY 1999 includes a study of skeletal development in embryonic quail. This study might eventuate in follow-up studies on the International Space Station. Also funded for FY 1999 are an assessment of the musculoskeletal effects of a growth hormone-releasing hormone agonist in unloaded mice and a study of the prevention of oxidative damage to bone during microgravity by vitamin E.

Human Studies

For human studies, the *Strategy* report recommended the development of a comprehensive description of microgravity-induced bone loss, using state-of-the art noninvasive methods entailing a careful record of skeletal changes postflight for each astronaut. It recommended initiation of a comprehensive database for the purposes of correlating skeletal changes with age, gender, muscle changes, diet, and genetic factors. To facilitate countermeasure development it recommended that the mechanisms of bone loss be defined using contemporary biochemical markers of bone turnover. Finally, it recommended that both loading and pharmacological interventions be pursued as potential countermeasures. To date, essentially no research has been conducted in accord with the first three of these recommendations. However, the opportunity to address the first two has been severely inhibited by operational constraints imposed by International Space Station (ISS) construction and by the limitation of missions to brief flights.

In human studies not included in *Strategy* report recommendations, ARC biomechanics program scientists have developed, validated, and initiated evaluation of a shoe insert that provides quantitative information about load cycles and cycle intensity over several consecutive days. This information should permit further elucidation of the relationships between habitual mechanical loading and bone mass and architecture. In addition, limited work at ARC continues with the human head-tilt bed rest model, but the high cost of this type of study may require external collaborations in the future.

Several new projects were funded for FY 1999. These include an evaluation of a skin patch for monitoring the loss of calcium in sweat and bone turnover. Validation of such a technique would greatly simplify the collection of in-flight data on mineral metabolism. Another project involves development of a dual energy X-ray system for bone mineral density and diagnostic radiography. At JSC, a project has been initiated to estimate the calcium kinetics and bone turnover on spaceflight missions. This project represents a continuation of work done in association with the Mir program.

To date, progress congruent with many of the *Strategy* report recommendations has been made in basic, animal, and human studies. Lack of progress in recommended areas appears to be largely due to limited opportunity for conducting flight experiments.

PROGRAMMATIC BALANCE

Balance of Subdiscipline Areas

The program as a whole contains an appropriate mixture of both basic and applied studies having clear relevance to *Strategy* report recommendations. Among current bone-related research projects, three (two human, one animal) specifically test potential countermeasures, two assess fundamental aspects of drug and hormone action on bone that could lead to new countermeasure development, two explore innovative technologies aimed at improved countermeasures, and one provides additional validation of an animal model.

Balance of Ground and Flight Investigations

The *Strategy* report recommended a series of studies to characterize the magnitude and regional distribution of bone loss during flight, as well as postflight changes. It also recommended the establishment of a systematic database to permit correlations of skeletal changes with a number of factors, including age, gender, muscle and hormonal changes, and diet. In addition, a series of mechanistic studies were recommended to give insight into the relative contributions of bone resorption and formation to the skeletal changes of microgravity. However, the paucity of flight opportunities currently requires the bone program to be entirely ground based, and except for continuation of a calcium turnover project, no work currently addresses these recommendations. Successful completion of several of the current studies would lead logically to flight-based investigations in the future.

Emphasis Given to Fundamental Mechanisms

The *Strategy* report recommended a series of experiments designed to understand fundamental mechanisms of bone loss. At ARC, an independent program of basic cellular and molecular biology has been initiated. In addition, work with hindlimb-unloaded rodent models has components addressing fundamental hormonal responses. One of four NSBRI-funded laboratory groups is conducting an extensive cell and molecular biology program addressing skeletal aspects of reproductive and vitamin D steroid receptor physiology. Several of the extramurally funded programs also explore changes in bone cell regulation, signal transduction, and the biochemistry of bone-specific proteins during unloading.

Utilization and Validation of Animal Models

The primary animal model used in investigations of bone loss and its mechanisms has been the hindlimb-unloaded rat. Over the past decade a large number of studies have been published that define and refine this model as well as demonstrate its value as a surrogate for space-based bone loss. The *Strategy* report recommended that the distribution and characteristics of bone loss throughout the skeleton be defined more completely and also that this model be the standard for comparing new models as they are introduced. The report specified that additional validational studies should involve bone histomorphometry, biomechanical and biochemical characteristics of unloaded bone, regulation of calciotropic hormones during hindlimb unloading, and features of cultured osteoprogenitor cells from unloaded animals. At present, little work has been done on these recommendations, but studies of this type are currently being initiated.

Overall, the bone program shows a reasonable distribution of effort among basic cellular, animal, and human studies. The paucity of flight opportunities currently requires the bone program to be entirely ground based. However, successful completion of several of the current studies would lead logically to flight-based investigations. In consonance with *Strategy* report recommendations, a strong effort is being conducted to address cellular and physiological mechanisms. Studies to address recommendations regarding animal model validation are being initiated. By permitting the conduct of transgene and gene knockout experiments, extension of hindlimb-unloading studies to mice will permit studies to assess genetic contributions to skeletal response. Further validation of the hindlimb-unloaded mouse should be conducted.

DEVELOPMENT AND VALIDATION OF COUNTERMEASURES

Most skeletal countermeasures that have been employed to date have involved some form of mechanical loading, such as treadmill exercise, with or without bungee cords, or resistance exercise. These generally have been unsuccessful. Probably the best hope for a successful intervention lies in a program that replicates or at least approaches normal gravitational force. Compatible with a *Strategy* report recommendation, two countermeasure projects have been conducted. An ARC program has developed a prototype positive pressure device that permits establishment of regional areas of increased "gravitational" loading for individuals in a microgravity environment. Results of a recent JSC study show that a combination of heavy resistance exercise plus a bisphosphonate can attenuate bone loss in humans subjected to bed rest. Effective collaborations are maintained with local universities to continue this work. An additional approach could involve administration of newly developed drugs and biological agents that regulate osteoblast or osteoclast function as a countermeasure against bone loss.

Despite the fact that countermeasure studies are in progress at ARC, JSC, and NSBRI laboratories, there is little evidence of coordination of these programs or, for that matter, of a system-wide coordinated process for testing and validating potential skeletal countermeasures as recommended by the *Strategy* report. NASA personnel at JSC seemed relatively unaware of ARC programs, and no formal mechanism exists to encourage bringing the results of these programs to program leaders at JSC.

EPIDEMIOLOGY AND MONITORING

Comprehensive monitoring of bone mineral density and turnover in flight crews was instituted in the past. However, despite success in obtaining sequential bone mass measurements and specimens for turnover measurements, the results of these studies are fragmentary and not generally available. The current Astronaut Medical Evaluation Requirements Document (AMERD) for missions exceeding 30 days mandates dual energy x-ray absorptiometry (DXA) assessment of body composition and bone mineral density one month prior to, and about one week after, completing the mission. The *Strategy* report clearly recommended the collection and maintenance of a database adequate to document and characterize the nature and distribution of bone loss during spaceflight missions. However, it does not appear that a strategy for systematic analysis and interpretation of these measurements has emerged. Similarly, repository specimens of urine and blood from previous missions could be used to generate important information regarding the time course and characteristics of altered bone remodeling during flight, but plans to use these specimens for such purposes are not described in the resource documents.

One problem created by the focus of current missions on Space Station construction is that the duration in microgravity for any given crew member will not exceed 30 days. Such flights would neither be covered by the requirement for DXA measurements nor be likely to provide insights relevant to long-term spaceflight. Thus, epidemiology and monitoring are currently at a standstill. However, there remains an opportunity to derive useful research information from even these short-term missions. Current practice includes collection of multiple urine and blood specimens pre- and postflight. Inclusion of in-flight data collections would provide substantial benefit, but this is not currently part of routine flight procedure. Assessment of bone turnover markers using pre-, intra-, and postflight specimens should afford valuable insights regarding the skeletal effects of flight and of countermeasure intervention. In this regard, by the terms of the ISS Medical Operations Requirements Document (MORD), in-flight health and fitness evaluations are routinely mandated to assess crew health and also to validate and adjust countermeasures (Section 3.6.5). Also mandated (Section 6.2.4.1) are individualized in-flight programs of resistive and aerobic exercise. Access to specimens and data remains problematic. Analysis of in-flight specimens for markers of bone resorption and formation would offer a unique opportunity to determine the relative efficacy of these various exercise programs.

The *Strategy* report recommended the collection of a comprehensive database concerning changes in bone mass and turnover during spaceflight. However, no strategy for analyzing, interpreting, and disseminating the results of flight measurements currently exists. Biological specimens from previous and current missions could be used to generate important information, but no plans are described for their use. Skeletal epidemiology and monitoring are currently dormant, but opportunities exist to derive useful information even from current short-term missions.

SUPPORT OF ADVANCED TECHNOLOGIES

The *Strategy* report recommended development of both general and bone-specific equipment. General equipment concerned modifications of animal facilities, animal centrifuges, and animal handling equipment to be used during spaceflight. Bone-specific equipment included modified bone densitometers for use in humans and animals during flight and exercise instruments to apply different types of mechanical stimulation (e.g., low- and high-impact loading). No studies currently address general instrumentation needs. Bone-specific instrumentation work is summarized below.

State-of-the art methodology is being applied in new ways. Application of contemporary molecular biology techniques to mechanically loaded cells, the load-sensing shoe insert, new computer algorithms for resolving bone geometry from DXA scans, and the miniature bone densitometer described above are examples of such advanced technology.

In accord with *Strategy* report recommendations and as part of an NSBRI Technology Development Program, an innovative miniature bone densitometer is currently under development that will theoretically be sufficiently portable for skeletal monitoring to be accomplished during the course of an extended mission. Moreover, it should be possible for such an instrument to measure body composition in addition to bone mass.

In accord with *Strategy* report recommendations for bone-specific instrumentation, the bone program appears to be providing an appropriate level of support for advanced technologies. General instrumentation needs are not currently being addressed.

SUMMARY

The current bone program is broad in scope and includes projects that could generate countermeasure interventions. The program appears to be appropriately balanced in terms of basic and applied science. Many of the specific projects are highly consistent with goals of the *Strategy* report recommendations. Other specific recommendations for both animal and human research have not yet been addressed. With respect to countermeasure development, one identified problem is the apparent lack of a formal mechanism to bring potential countermeasures to formal assessment, as well as an apparently haphazard and unfocused epidemiology and monitoring component that fails to take advantage of unique and potentially valuable research data.

BIBLIOGRAPHY

National Aeronautics and Space Administration (NASA). 1997. Task Force Report on Countermeasures: Final Report. Washington, D.C.: NASA.

NASA. 1998a. Life Sciences Program Tasks and Bibliography for FY 1998. Washington, D.C.: NASA.

NASA. 1998b. International Space Station Medical Operations Requirements Document (ISS MORD), Baseline SSP 50260. Houston, Tex.: NASA.

NASA and Universities Space Research Association (USRA). 1999. Proceedings of First Biennial Biomedical Investigators' Workshop, January 11-13, 1999, League City, Texas. Houston, Tex.: NASA and USRA.

National Research Council (NRC), Space Studies Board. 1998. A Strategy for Research in Space Biology and Medicine in the New Century. Washington, D.C.: National Academy Press.

4

Muscle Physiology

INTRODUCTION

Human spaceflight results in loss of skeletal muscle mass, diminished strength and endurance, degraded motor performance, and increased susceptibility to reloading injury. The in-flight changes begin within a few days and increase in severity with time in space. Exercise countermeasures are marginally effective. Muscle deterioration remains a significant crew health and safety issue with a high potential to adversely affect missions, especially those involving extravehicular activities (EVAs) and transitions into a gravity environment such as Mars, as well as emergency egress. For example, the committee was informed by NASA that assisted egress was often required after long-duration missions. For emergency egress, astronauts need to be as fit as possible, and this includes both strength and aerobic capacity. Although prolonged intensive use of leg muscles may not be necessary in flight, it will be required on reentry and for future planetary exploration.

Maintaining muscle health requires elucidating underlying mechanisms and developing effective countermeasures through well-controlled, scientific flight investigations. Extensive attempts to obtain such data over the past decade through supplemental mission objectives rather than as primary investigations have not yielded definitive results (NASA, 1987, 1991, 1994, 1999). The absence of a formal process for countermeasure evaluation open to extramural input and the paucity of flight opportunities hamper in-flight validation of potential countermeasures cultivated in ground-based experiments. The in-flight exercise employed to date has afforded little protection against muscle wasting. Consensus is lacking on the specific exercise protocols required to maintain fitness, and the absence of a definition of fitness further compounds the problem.

The *Strategy* report (NRC, 1998) recommended that priority be given to research focusing on cellular and molecular mechanisms underlying muscle deterioration. Human and animal ground-based simulations should be exploited to investigate spaceflight effects on muscle mass, protein composition, myogenesis, fiber-type differentiation, and neuromuscular development. It is important to determine how muscle cells sense working length and the mechanical stress of gravity. The recommended approaches include analysis of signal transduction pathways for growth factors, stretch-activated ion channels, regulators of protein synthesis, and interactions of extracellular matrix and membrane proteins with the cytoskeleton.

NASA'S CURRENT RESEARCH PROGRAM IN MUSCLE PHYSIOLOGY

The total expenditure for the Biomedical Research and Countermeasure (BR&C) program in FY 1999 was about $2.805 million distributed among approximately 18 projects in the NASA Research Announcement (NRA) and National Space Biomedical Research Institute (NSBRI) programs (see Table 4.1). Current funding is distributed as approximately 60 percent for human studies and 40 percent for animal studies. About nine investigations with at least a partial emphasis on muscle physiology are supported by the NRA process ($1.801 million), and there are six main projects in the NSBRI program ($1.087 million).

The human studies focus primarily on countermeasures, including the utility of lower-body negative pressure (LBNP) and resistance exercise. Three human flight protocols funded in FY 1999 involved a pre-post muscle biopsy study, a test of foot function during spaceflight, and an examination of protein turnover during spaceflight.

Investigators in NASA centers, NSBRI, and NRA extramural programs collaborate on exercise testing during bed rest, rodent hindlimb unloading (exercise, pharmacological and hormonal countermeasures), calcium signaling in human skeletal muscle cultures, and gravity effects on sarcolemmal structure and function. Parabolic flights producing brief periods of reduced gravity are used to probe rapid changes in muscle cell membranes.

The First Biennial Space Biomedical Investigators' Workshop in 1999 demonstrated that excellent mechanistic research is being conducted on skeletal muscle that is in line with the 1998 *Strategy* report recommendations (NASA and USRA, 1999; NRC, 1998). The presentations included fundamental studies on fiber-type differentiation, multiple signaling pathways, hormonal regulation, and control of fiber-type gene expression. Investigators pursued novel genetic and biochemical strategies for combating muscle atrophy during rodent hindlimb unloading, such as growth hormone (GH)/insulin-like growth factor-I (IGF-I) gene therapy in genetically engineered mice and pharmacologic blockage of the ubiquitin degradation pathway. The inclusion of genetically modified organisms in the muscle research program is consistent with the *Strategy* report recommendation to exploit these animal models.

The Fundamental Biology Research Program (FBRP) also supports through the NRA process muscle physiology studies that explore fundamental mechanisms. Five of these investigations probe the signaling of muscle atrophy, skeletal muscle artery adaptation, muscle growth and repair, insulin signal

TABLE 4.1 Summary of Funding in FY 1999 for Muscle Physiology Subdisciplines

Subdiscipline	NRA		NSBRI	
	Total ($ thousands)	No. of Projects	Total ($ thousands)	No. of Projects
Organism				
Ground	758	4	40	1
Flight	750	2	—	—
Countermeasure	293	3	—	—
Cell and molecular	—	—	1,047	8
Total	1,801	9	1,087	9

transduction in muscle, and mechanical signal transduction in countermeasures to muscle atrophy in rats subjected to hindlimb suspension unloading. Two of the studies involve muscle cultures to explore growth factors and tension-induced skeletal muscle growth and the effects of space travel on skeletal myofibers.

In summary, research to date has been descriptive, and the next generation of experiments promises to elucidate mechanisms and identify effective countermeasures. The objectives of the muscle physiology investigations supported by NASA's BR&C and FBRP are consistent with the *Strategy* **report recommendation to determine how muscle cells sense the mechanical stress of gravity. In agreement with the** *Strategy* **report, continuing ground and flight studies of human and animals are necessary to advance the understanding of debilitation and maladaptation of muscle to spaceflight.**

PROGRAMMATIC BALANCE

Balance of Subdiscipline Areas

NASA muscle research ranges from basic studies of cells to studies of the whole organism. A portion of the funding is directed at countermeasure testing in human ground-based models. The collection of investigations represents an appropriate balance of cellular, molecular, and whole-animal mechanistic studies and human countermeasure evaluation.

Balance of Ground and Flight Investigations

There have been many excellent flight investigations, most recently on Spacelabs SLS-2 and Neurolab and on earlier Spacelab missions and rodent experiments carried in the Shuttle middeck. **However, with closure of the Spacelab era, Mir-Shuttle missions, and the Bion program, flight opportunities have largely disappeared. Investigations will continue on Spacehab and Shuttle missions, but they will be much reduced in number and scope because the prime objective of these missions is currently International Space Station (ISS) construction rather than experimentation. Ground-based research will, therefore, necessarily predominate until ISS completion. Unfortunately, this hiatus in flight investigations further delays testing, validation, and implementation of countermeasures to ameliorate muscle wasting.**

Emphasis Given to Fundamental Mechanisms

High priority is appropriately assigned to research on cellular and molecular mechanisms, utilizing ground-based models for human and animal studies. Most of the NSBRI-funded animal work pursues basic mechanisms. This emphasis agrees with the *Strategy* report recommendation to employ such models for testing and refining hypotheses seeking to understand the fundamental mechanisms of how workload is transduced into molecular signals regulating muscle properties. The muscle research program focuses on (a) elucidating the cellular and molecular mechanisms underlying muscle atrophy, weakness, faulty coordination, and delayed muscle soreness, (b) the use of appropriate ground-based models to investigate these problems, and (c) the process by which muscle senses working length and the mechanical stress of gravity. This emphasis is compatible with the *Strategy* report recommendations.

Utilization and Validation of Ground and Animal Models

Continued research on muscle changes during human bed rest and rodent hindlimb unloading is justified for investigating basic mechanisms. These models simulate spaceflight changes very well for skeletal muscle. During bed rest, muscle responds positively to resistance training by partial amelioration of atrophy (Bamman et al., 1998). Given the limited access to spaceflight and low numbers of flight subjects, selection of muscle wasting countermeasures for flight will rely increasingly on results of well-controlled studies in ground-based models. The rodent model of hindlimb unloading, traditionally used for rats, also works well with normal and genetically altered mice. This provides valuable breadth to the investigations of mechanism, a point emphasized in the *Strategy* report. For humans, a buoyancy-equipped treadmill at the Ames Research Center (ARC) simulates the effects on locomotion at a reduced loading (0.38 g) environment of Mars. The centrifuge research facilities at ARC are serving the renewed interest in hypergravity investigations generated by the projected use of acceleration to simulate gravity on ISS and the potential use of artificial gravity as a countermeasure.

Congruent with the *Strategy* report, the optimal balance of studies necessitates greater reliance on ground-based models, mostly involving animal studies, to elucidate basic mechanisms, refine hypotheses, and assess the efficacy of countermeasures because of the paucity of flight experiment opportunities.

DEVELOPMENT AND VALIDATION OF COUNTERMEASURES

The Extended Duration Orbiter Medical Project (EDOMP) succeeded the Detailed Supplemental Objective (DSO) Program when Shuttle flights became 16 days or longer (NASA, 1999). EDOMP continued the testing of countermeasures as secondary mission objectives with the goal of optimizing crew ability to maintain performance. As part of the EDOMP, in-flight aerobic exercise was proposed for evaluation. Appropriate data were sought, but the amounts and quality obtained were insufficient to reach unambiguous conclusions. In-flight cycle ergometry exercise ameliorated cardiovascular aerobic deconditioning but not skeletal muscle deterioration. Maximal exercise testing on the treadmill on landing day (DSO 476) was abandoned in 1988 because the test produced delayed-onset muscle soreness, whereas cycle ergometers did not have this effect (NASA, 1999). Currently, postflight testing is limited to minimizing risk of muscle injury. This avoidance philosophy leaves the muscle reloading injury problem unresolved. Countermeasures utilizing resistive lengthening muscle contractions hold promise, but their efficacy must be assessed during postflight transition to gravity loading. By and large, the supplemental objectives have not provided definitive answers to the muscle debilitation and countermeasure problem. The DSOs were specifically designed not to interfere with primary mission objectives, and unlike scientific investigations with fixed aims, the objectives of DSOs were changed frequently in response to the needs of the flight program. Exercise protocols are tailored to individuals. The lack of standard protocols across subjects greatly hampers distinguishing individual response from protocol variation. Following the Challenger accident, the exercise protocols emphasized the new requirement for unaided emergency egress. The uncontrolled nature of these studies and the small number of subjects contributed to the failed consensus on defining an optimal exercise regimen for maintaining muscle health.

Medical Operations at the Johnson Space Center (JSC) directs muscle countermeasure studies for humans and regulates the repertoire of measures tested. Valuable input from extramural labs is lacking. This constrains the breadth of countermeasure development. A formal process is necessary to capitalize on the novel countermeasures expected to emerge from basic cellular and molecular research, a major focus of the NSBRI Muscle Atrophy Team (NSBRI, 1998). Hypo- and hyperloading on treadmills, under investigation at ARC, represent promising countermeasures (NASA, 1997a). Although JSC is developing elaborate plans for executing countermeasure studies, the countermeasure program suffers from the lack of a formal mechanism for transitioning potential countermeasures from ground-based investigations to flight testing (NASA, 1998a). The current database is inadequate to define a strategy for meeting the goal of maintaining muscle health during and after long-term spaceflight.

Maintenance of muscle health requires a multipronged approach. The *Task Force Report on Countermeasures* (NASA, 1997b) concluded that existing cycling, rowing, and treadmill exercise protocols did not maintain muscle mass and a positive nitrogen balance. However, the benefits of muscle stretching during these exercises cannot be overlooked. A side benefit of aerobic conditioning exercise is that leg muscles are stretched nearly through their full range. Ground-based studies have shown that muscle length regulation depends on the working range of movement. In these exercises, loading forces were insufficient to conserve muscle mass and prevent conversion from slow to fast isomyosin. The report recommends high priority for resistance exercise training and the use of existing exercise hardware in the short term. Appropriate countermeasures should maintain the mass of both skeletal and cardiac muscle and preserve motor skills and posture. Fitness standards, skill levels, and performance criteria should be defined. Countermeasure program design should foster compliance and ensure adequate nutrition. The Exercise Physiology Laboratory and Muscle Research Laboratories at JSC study mechanical loading and growth factor release as potential countermeasures for preserving muscle mass. There are pending ground studies of Russian exercise countermeasures and new resistance exercise countermeasures. As concluded in the *Strategy* report, LBNP coupled with resistance exercise should be tested for maintaining the microcirculation and strength of the muscle. NASA funded an LNBP investigation in 1999 as an advanced technology development countermeasure.

A current NASA flight rule requires that some exercise take place on all missions of 11 days or longer duration. Current exercise regimens do not preserve sufficient muscle strength and are unlikely to prevent susceptibility to reloading injury. Countermeasure testing and validation targeting operationally relevant activities must be primary objectives on future spaceflight missions. Without this change or a formal process to translate potential countermeasures from both intra- and extramural laboratories, progress in solving muscle-related problems will fall further behind that required to support the continuous presence of humans in space.

EPIDEMIOLOGY AND MONITORING

Plans for Monitoring Crew Health and Fitness on the International Space Station

A program of operational monitoring and validation of in-flight countermeasures will be implemented on the ISS (NASA, 1998a). Additional countermeasures may be added based on perceived need and scientific verification. Currently, mission length determines countermeasure requirements. The battery of proposed in-flight countermeasures includes exercise (treadmill, required cycle ergometer and resistive exercise, LBNP, and electromyostimulation). Although these interventions may have merit, the rationale for their inclusion is not solidly based on proven effectiveness for maintaining muscle health.

The Environmental Physiology and Biophysics Laboratory will examine exercise intensity during extravehicular activity and monitor its effect on decompression sickness and muscle deconditioning. The Nutritional Biochemistry Laboratory tracks nutritional intake on selected flights and cooperates with operational medicine in deciding on individualized menus for flight.

International collaboration is evident in the Russian Motomir project to develop resistance exercise devices, and interactions utilizing the European Space Agency (ESA) Muscle Atrophy Research and Exercise System (MARES) in bed rest experiments. This hardware is slated for use on ISS. The International Space Life Sciences Working Group should coordinate these efforts. In the era of ISS, progress on preventing muscle debilitation will depend greatly on close coordination of medical operations monitoring of crew health and fitness with basic research flight investigations on humans. Assessment of the health of human skeletal muscle is proceeding on a variety of fronts, such as measuring fatigue and high-energy phosphates by magnetic resonance imaging (MRI) and spectroscopy.

Increased attention to noninvasive monitoring of muscle function and health is in accord with the *Strategy* report recommendation to document more thoroughly the history of muscle use for astronauts.

SUPPORT OF ADVANCED TECHNOLOGIES

In the 1998 NRA, the BR&C program called for advanced technology development of implantable sensors, electronics, and software to monitor and analyze long-term nerve and muscle activities in unrestrained subjects as well as exercise apparatuses, which provide a means of quantifying the amount and pattern of work performed while controlling the levels and patterns of loading imposed on muscles, bones, and joints (NASA, 1998b). Documenting individual activity in flight by automated recording unburdens crew members from manual logging tasks. Moving in this direction is consistent with the *Strategy* report recommendation for improved documentation of individual activity before, during, and after flight that would greatly facilitate interpretation of results in muscle investigations.

The Advanced Technology Development Division at ARC is perfecting quantitative computed tomography and simulated altered gravity locomotion under terrestrial conditions, using LBNP, upper-body positive pressure, and lower-body positive pressure. The Sensors 2000 program at ARC interacts with the NSBRI Technology Program to advance telemetric-based sensor systems for measuring muscle properties such as muscle activity, interstitial pressure, and blood flow. The goal is noninvasive or mildly invasive physiological monitoring of astronauts, utilizing instruments to display health status in flight.

Additional advanced technologies are being pursued by NSBRI, NASA centers, and NRA extramural technology development programs. These include development of rapid freeze equipment to preserve muscle cell constituents for biochemical and molecular analyses, more ergonomic space suits to reduce muscle fatigue, and imaging instruments for documenting muscle deterioration and real-time assessment of countermeasure efficacy. Closer interaction between muscle researchers and engineers designing space suits is desirable. Coordination of the efforts of these programs promises to increase advances in muscle physiology.

In agreement with the *Strategy* report, the development of new technologies for noninvasive monitoring of astronaut muscle physiology is being pursued with the goal of improved definition of the history of muscle use for clearer interpretation of the effects of spaceflight and the efficacy of countermeasures.

SUMMARY

Skeletal muscle deterioration remains a significant crew health, performance, and safety issue for spaceflight. Research to date has identified exercise as affording minimal protection. However, the *Strategy* report and the Countermeasures Task Force report point out that the lack of well-controlled, scientific flight investigations prevents consensus on specific protocols for maintaining fitness. The *Strategy* report emphasizes that the paucity of flight experiment opportunities necessitates greater reliance on ground-based models for refining hypotheses and assessing countermeasure efficacy and implementation. In the era of ISS, progress on preventing muscle debilitation will depend greatly on the close coordination of medical operations monitoring of crew health and fitness with basic research flight investigations on humans. New technologies for noninvasive monitoring of muscle health during spaceflight are being developed. This should achieve the *Strategy* report goal of improved documentation of individual astronaut's history of muscle use to control for intersubject variation and reduce uncontrolled variables. In the past, countermeasure testing and validation were not the primary objectives of spaceflight missions. Without manifesting them as high-priority goals, as recommended in the *Strategy* report, and instituting a formal process to incorporate potential countermeasures from both intra- and extramural laboratories, progress in solving muscle-related problems will fall further behind that required to support the continuous presence of humans in space. Congruent with the *Strategy* report, the muscle research program is putting great effort into understanding the basic cellular and molecular mechanisms for atrophy, weakness, and susceptibility to injury. Progress is being to be made in defining how muscle cells sense working length and load imposed by gravity through unloading and hypergravity ground-based studies of normal and genetically altered rodents, as recommended by the *Strategy* report. These studies are exploring hormones, growth factors, second messengers, and drugs that potentially translate into novel countermeasure applications. Additional studies called out in the *Strategy* report are needed to examine nerve and muscle repair, failed microcirculation, and fundamental aspects of myogenesis and regeneration because muscle damage and repair during spaceflight are inevitable in spite of rigorous safety practices and countermeasures.

REFERENCES

Bamman, M.M., M.S. Clarke, D.L. Feeback, R.J. Talmadge, B.R. Stevens, S.A. Lieberman, and M.C.J. Greenisen. 1998. Impact of resistance exercise during bed rest on skeletal muscle sarcopenia and myosin isoform distribution. Appl. Physiol. 84(1):157-163.

National Aeronautics and Space Administration (NASA). 1987. Results of the Life Sciences DSOs Conducted Aboard the Space Shuttle 1981-1986. M.W. Bungo, T.M. Bagian, M.A. Bowman, and B.M. Levitan, eds. Houston, Tex.: NASA.

NASA. 1991. Results of Life Sciences DSOs conducted Aboard the Space Shuttle 1988-1990. Houston, Tex.: NASA.

NASA. 1994. Results of Life Sciences DSOs Conducted Aboard the Shuttle 1991-1993. Houston, Tex.: NASA.

NASA. 1997a. Life Sciences Division Report. NASA Ames Research Center. Moffett Field, Calif.: NASA.

NASA. 1997b. Task Force Report on Countermeasures: Final Report. Washington, D.C.: NASA.

NASA. 1998a. International Space Station Medical Operations Requirements Document (ISS MORD), Baseline SSP 50260. Houston, Tex.: NASA.

NASA. 1998b. NASA Research Announcement: Space Life Science—Research Opportunities in the Advanced Support Technology (AHST) Programs. NRA-98-HEDS-01. Washington, D.C.: NASA.

NASA. 1999. Extended Duration Orbiter Medical Project Final Report 1989-1995. C.F. Sawin, G.R. Taylor, and W.L Smith, eds. NASA SP-534. Houston, Tex.: NASA.

NASA and Universities Space Research Association (USRA). 1999. Proceedings of the First Biennial Biomedical Investigators' Workshop, January 11-13, 1999, League City, Texas. Houston, Tex.: NASA and USRA.

National Research Council (NRC), Space Studies Board. 1998. A Strategy for Research in Space Biology and Medicine in the New Century. Washington, D.C.: National Academy Press.

National Space Biomedical Research Institute (NSBRI). 1998. Annual Report: October 1, 1997-September 30, 1998. Houston Tex.: NSBRI.

5

Cardiovascular and Pulmonary Systems

INTRODUCTION

The cardiovascular (CV) system is tightly regulated, gravity dependent, and extremely adaptable. The plasticity of this system results in a rapid, but not immediate, adjustment to weightlessness that occurs within the first few days of spaceflight. However, the system is unable to reverse its adaptation to spaceflight immediately upon return to normal gravity, and this delay leads to the physiologically and operationally important limitations in CV function that have been observed with reentry into Earth's gravitational field. In addition to postflight orthostatic intolerance, the main CV issues of potential importance for spaceflight include in-flight aerobic deconditioning, cardiac atrophy, cardiac arrhythmias, and long-term effects on crew member health of the CV adaptive changes induced by spaceflight. The major concerns with respect to the pulmonary system, and the main foci of pulmonary research, are adequate denitrogenation prior to extravehicular activities (EVAs), increased aerosol deposition in spaceflight, and changes in pulmonary perfusion.

Both the CV and the pulmonary systems have been studied extensively in dedicated life science flights as well as experiments performed for Detailed Supplemental Objectives (DSOs) and in the Extended Duration Orbiter Medical Project (EDOMP) program. Nonetheless, orthostatic intolerance remains an operational problem, and fundamental questions remain about the nature, degree, and severity of spaceflight-related cardiac atrophy and arrhythmias. Differences of opinion and confusion exist concerning the incidence of arrhythmias during and following spaceflight. The *Strategy* report (NRC, 1998) did not identify arrhythmias as a significant concern, whereas the National Space Biomedical Research Institute (NSBRI) and the Johnson Space Center (JSC) have targeted the potential for arrhythmias as a major concern. With respect to cardiac atrophy, there was a report from the postflight workshop for the Deutsche 2 Spacelab Mission (Blomqvist, 1995) indicating significant cardiac atrophy during this two-week flight. Although one data set does not constitute clear and convincing evidence, the problem of cardiac atrophy is potentially significant enough to require follow-up and confirmation. For the pulmonary system, an excellent basic physiologic understanding has been established, but operational problems, such as prebreathing protocols and aerosol deposition, remain.

The *Strategy* report made several recommendations in the cardiovascular area. These recommendations were to determine the following: (a) adaptive responses of the cardiovascular system to spaceflight and the mechanisms underlying these adaptations, (b) immediate postflight cardiovascular responses and their mechanisms, (c) adequacy of ground-based models to replicate spaceflight-induced cardiovascular changes, (d) long-term consequences of extended spaceflight on cardiovascular health, and (e) the need for and validation of countermeasures to be applied for the cardiovascular effects of spaceflight. The major pulmonary recommendations were to determine the following: (a) effects of microgravity on aerosol deposition, including the possible effects of lunar or Martian dust particles, (b) effects of long-duration spaceflight on pulmonary and respiratory muscle function, and (c) optimal denitrogenation protocols for Space Station EVA. Technology needs identified in the *Strategy* report were for (a) advanced physiological data systems (e.g., heart rate, beat-to-beat blood pressure, cardiac output, gas exchange); (b) improved imaging systems (e.g., scintigraphic system for spaceflight use); (c) new exercise equipment; (d) advanced aerosol monitoring systems; and (e) improved decompression sickness monitoring equipment.

The following sections summarize the findings relevant to these recommendations.

NASA'S CURRENT RESEARCH PROGRAM IN CARDIOVASCULAR AND PULMONARY SYSTEMS

At an FY 1999 funding level of about $4.9 million, CV studies account for approximately 14 percent of the total NASA budget for biomedical and countermeasure research funding. Additional funding comes from sources such as operations (e.g., to support the clinical evaluation of reduced prebreathe times and for integrated testing regimes), but the magnitude is unknown. The NASA Research Announcement (NRA) and NSBRI listings indicate a total of 29 projects in the cardiopulmonary discipline. Twenty-seven studies are performed either entirely or partially in humans. Twenty studies address CV physiology and/or orthostatic changes, two address arrhythmias, three focus on the pulmonary system, one focuses on atrophy, and three are concerned primarily with countermeasure evaluation. Table 5.1 summarizes the projects and funding.

Cardiopulmonary projects are carried out at JSC, the Ames Research Center (ARC), NSBRI-funded university laboratories, and other university laboratories through extramural grants. The Environmental Physiology Laboratory at JSC supports studies on EVA, and studies on pulmonary function are supported through NRA and flight projects programs at extramural laboratories. The Medical Operations

TABLE 5.1 Summary of FY 1999 Funding for Cardiovascular and Pulmonary Systems Subdisciplines

Subdiscipline	NRA		NSBRI	
	Total ($)	No. of Projects	Total ($)	No. of Projects
Pulmonary	596,000	3	0	0
Orthostatic intolerance	2,754,195	14	549,482	6
Cardiac atrophy	0	0	143,050	1
Cardiac arrhythmias	0	0	247,087	2
Countermeasures	662,000	3	0	0
Total	4,012,195	20	939,619	9

Section collects CV and pulmonary data on astronauts as part of the physical certification process and for the longitudinal study of astronaut health.

Currently, major inconsistencies exist among the programs (NRA, NSBRI, operational medicine) in the perception of the relative importance of the various cardiovascular changes that may result from spaceflight. To resolve these differences, it is recommended that a "cardiovascular summit" be convened to review all existing cardiovascular data from spaceflight, ground-based studies, and animal studies. Summit attendees should include representatives from the NSBRI, NASA operational medicine, NASA intramural investigators, the Astronaut Office, and outside experts in cardiovascular medicine and physiology. The charge of the group would be to:

- **Determine if cardiac arrhythmias are a significant concern and if monitoring is warranted;**
- **Determine if cardiac atrophy is a significant concern and should be monitored;**
- **Determine if there are pre- and in-flight predictors of orthostatic intolerance;**
- **Determine which ground-based human and animal models best reproduce the effects of spaceflight;**
- **Review the data and rationale for current countermeasures and suggest new countermeasures on the basis of the existing physiological data;**
- **Determine what level of aerobic fitness should be maintained in space; and**
- **Determine if countermeasures have been effective when used as recommended.**

PROGRAMMATIC BALANCE

Balance of Subdiscipline Areas

The present complement of CV experiments shows appropriate balance between studies of orthostatic intolerance, arrhythmias, and basic physiological mechanisms, but human studies on cardiac atrophy are lacking. The pulmonary investigations are heavily focused on denitrogenation protocols and decompression sickness. There appear to be no current studies on aerosol deposition, Martian dust effects, or respiratory muscle function as recommended in the *Strategy* report.

Balance of Ground and Flight Investigations

The CV and pulmonary programs are mainly ground based. The recent Neurolab mission had a significant payload of experiments aimed at studying CV autonomic regulation, but future missions of this sort are not currently planned. The bulk of current NRA- and NSBRI-supported investigations are ground based. The CV lab at JSC is involved in testing crews after Shuttle flights.

Future flight investigations will be subject to the Integrated Testing Regimen outlined in the Countermeasure Evaluation and Validation Project Plan (NASA, 1999b). The CV tests planned include tilt tests both before and after flights and exercise tests before, during, and after flights. The Countermeasure Evaluation and Validation Project Plan also mentions continuous cardiac rhythm (Holter) monitoring in flight and upon reentry, but both of these tests are currently canceled. No method to assess orthostatic responses in flight (i.e., lower-body negative pressure) is mentioned in the Integrated Testing Regimen (ITR).

In the area of EVA physiology, the use of shortened prebreathe protocols is being evaluated in a multicenter trial. The use of exercise to shorten prebreathe times and the use of argon-oxygen mixtures are under evaluation as part of the NRA program.

The current cardiovascular program is mainly ground based and thus is consistent with the *Strategy* report recommendation to do precursor ground-based studies. Future investigations to address the effects of spaceflight on the cardiovascular system will require human flight studies, as also recommended in the *Strategy* report.

Emphasis Given to Fundamental Mechanisms

Studies on basic physiological mechanisms receive appropriate priority. Several ongoing studies address the CV effects of spaceflight. These include NSBRI studies of embryonic development, microvasculature and tissue perfusion, molecular and cellular mechanisms of cardiac atrophy, and changes in gene expression. Since CV adaptations to spaceflight may depend heavily on integrated responses in the intact organism, understanding of this system is likely to be derived largely from physiological studies on intact animals or humans. The in vitro molecular and cellular studies under way should provide insight into the adaptations of the integrated CV system to spaceflight. **The current balance of molecular and cellular studies versus physiological studies appears consistent with the *Strategy* report.**

Utilization and Validation of Animal Models

There are several animal models under study, and a considerable volume of data is available from these model systems. As mentioned previously, there is a need to synthesize these data and develop a consensus as to which animal models best simulate different aspects of CV adaptation to microgravity in man. Various aspects of CV adaptation, such as atrophy, orthostatic hypotension, and arrhythmias, will likely require the use of different animal models to allow optimization of ground-based studies.

The current research program supports studies in rodents of cardiac atrophy, and the NSBRI also has a project on rodent studies of CV deconditioning. The project descriptions mention no specific plans for validating these models.

Plans for validating animal models are not evident and should be addressed by the Cardiovascular Summit.

DEVELOPMENT AND VALIDATION OF COUNTERMEASURES

A number of countermeasures against orthostatic intolerance either were previously utilized and abandoned (or inconsistently applied), are currently being utilized, or have been proposed but not yet tested. Few, if any, countermeasures have been proposed to protect against arrhythmias or cardiac atrophy. Further documentation is needed of the extent to which arrhythmias and atrophy actually constitute a problem before a decision is made to proceed with countermeasure development for these indications. **The Countermeasure Evaluation and Validation Project Plan should be utilized to evaluate countermeasures that are currently being deployed** as well as those proposed but not yet used. As new information becomes available, **consideration should be given to reevaluating selected countermeasures that have been previously abandoned to ensure that potentially useful approaches have not been discarded prematurely.** A partial list of countermeasures that have been proposed, (and in some cases are currently used but not adequately validated) includes the following: clothing (e.g., G-suit, elasticized (Penguin) suit, and liquid cooling garments); lower body negative pressure (LBNP) either alone or coupled with exercise; isotonic saline; supine reentry seat; pre-reentry maximal physical effort; resistive exercise; medications such as mineralocorticoid (Fluorinef) or adrenergic agents;

nutrients; periodic carotid baroreceptor negative pressure; continuous artificial gravity; and intermittent artificial gravity.

NASA initially supported a number of studies aimed at documenting CV adaptations to spaceflight. These studies have matured into experiments designed to understand the pathophysiology of the observed adaptations. The results of these mechanistic studies are beginning to suggest possible countermeasures based on confirmed physiological mechanisms. The Neurolab mission included detailed measurements of autonomic nervous system function in space that should demonstrate whether cardiovascular autonomic nervous system function is impaired in space. The results will help direct the efforts for countermeasure studies.

Of the active projects on the NRA and NSBRI lists, only three evaluate countermeasures. A multicenter trial of shortened prebreathe protocols has been undertaken to allow for shorter prebreath times on the International Space Station (ISS), but this is not supported by either the NRA process or the NSBRI. Exercise is also being studied as a way to reduce prebreathing requirements. In the NSBRI, none of the CV studies appears to be a countermeasure evaluation. The current fluid loading countermeasure appears not to be receiving ongoing evaluation, as evidenced by the cancellation of tilt testing after short-duration missions.

Only one new CV countermeasure (midodrine, an alpha-adrenergic agonist for orthostatic hypotension) is mentioned in the Countermeasure Evaluation and Validation Project Plan. This countermeasure does not appear to be part of an NRA or NSBRI project. Another countermeasure aimed at increasing blood volume (erythropoietin) did not receive NRA funding.

Few of the current cardiovascular studies evaluate current countermeasures or propose new ones. The appropriate testing regimens and candidate countermeasures should be a focus of the Cardiovascular Summit. The development of the Countermeasure Evaluation and Validation Project Plan and Integrated Testing Regimens addresses the *Strategy* report recommendation to validate new countermeasures.

EPIDEMIOLOGY AND MONITORING

Although considerable information has been gathered on CV function of crew members prior to, during, and following spaceflight, no formal program appears to be in place to monitor and review these data. Thus, these data have apparently not been synthesized into a comprehensive document available for scientific review. In addition, the most significant pathophysiologic changes noted thus far (orthostatic hypotension, cardiac arrhythmias, cardiac atrophy) have been studied to differing degrees in biomedical research flights but are not routinely evaluated using state-of-the-art techniques. It should be noted that although there are significant differences of opinion concerning the incidence of arrhythmias induced by spaceflight, no plans exist for routine cardiac rhythm monitoring. Some of the measurements under evaluation have great potential to assess cardiac arrhythmias and could be applied in a systematic way prior to, during, and after spaceflight. Measurement of cardiac mass by magnetic resonance imaging (MRI) is likely to give an accurate assessment of this potential issue, and consideration should be given to evaluating this methodology in comparison to echocardiographic methods currently in use. If MRI proves a more accurate means to detect loss of cardiac mass, it could be used pre- and postflight to determine the presence and tempo of resolution of this abnormality.

Integrated testing regimes have been outlined to make certain routine measurements on all flights. **The rationale for the tests included in these regimens is not documented and should be made explicit. A formal program should be in place for ongoing assessment of data collected on current or future spaceflights.**

The current monitoring program contains the following elements.

Orthostatic Intolerance

For short-term flights (<30 days) the operational tilt test is used only for first-time flyers or those with demonstrated functional orthostatic impairment. Functional orthostatic impairment is not defined in the Astronaut Medical Evaluation Requirements Document (AMERD; NASA, 1998a). For longer flights, the operational tilt test is mandatory. No in-flight assessment of orthostatic function is planned, so there will be no monitoring of orthostatic function prior to reentry. Reentry monitoring was planned for flights on the ISS but was canceled. It is recommended that in-flight monitoring (e.g., using LBNP) be added to assess orthostatic intolerance before reentry. Reentry monitoring would allow for orthostatic responses to be measured at the most operationally important time.

Cardiac Atrophy

Echocardiographic measurements will be taken during the operational tilt test. The AMERD does not mention if changes in cardiac mass will be monitored using this technology. In addition, no MRI measurements of cardiac mass are planned. **No program for monitoring cardiac atrophy exists in humans in the current program and the Cardiovascular Summit should address whether one is needed.**

Arrhythmias

Plans for cardiac rhythm monitoring on ISS missions were canceled. In flight, the electrocardiogram (EKG) will be monitored during EVA, and a 12-lead EKG will be assessed every 60 days. Every 30 days in flight, a physical assessment and submaximal exercise test of aerobic capacity will be conducted. How the results of these tests will be analyzed and used is not outlined in current documents, nor is the underlying rationale for the testing protocol.

The AMERD lists a submaximal exercise test to be performed every 30 days in flight. The rationale for this testing is not presented. Whether the crew needs to maintain a level of aerobic fitness is not explicitly stated.

Pulmonary

Pulmonary spirometric and peak flow measurements are planned for every 30 days in flight. An in-suit Doppler system has been developed for intravascular bubble monitoring during EVA, as recommended in the *Strategy* report. A system for aerosol generation and monitoring, recommended in the *Strategy* report, is not mentioned in the NRA or NSBRI programs.

The epidemiology and monitoring results are not integrated with the research program. It is recommended that a mechanism be developed to bring together representatives from the NSBRI, NASA operational medicine, NASA intramural investigators, and outside experts in cardiovascular medicine and physiology on an ongoing basis to review new data to assist in refining testing regimens and countermeasures. In the future, this might be a role for the integrated product teams (IPTs).

SUPPORT OF ADVANCED TECHNOLOGIES

Progress is being made to obtain sophisticated cardiac rhythm information prior to, during, and after spaceflight. One NSBRI program is developing CV system identification technology for measuring changes in autonomic nervous function. This work addresses the *Strategy* report recommendations to provide accurate measurements of heart rate and blood pressure. Two NRA projects involve exercise devices, and an in-suit bubble detection system is being used for intravascular bubble detection. Both of these efforts respond to recommendations in the *Strategy* report. Other *Strategy* report equipment recommendations (automatic physiological recording devices, cardiac output devices, gas analyzers for pulmonary measurements, scintigraphic imaging systems, aerosol-monitoring equipment) are not being addressed.

SUMMARY

Important questions, such as whether cardiac arrhythmias (or a propensity to develop arrhythmias) are stimulated by exposure to spaceflight and whether cardiac atrophy is significant and reversible, have not yet been answered. Mechanisms responsible for orthostatic intolerance, arrhythmias, and cardiac atrophy remain to be elucidated. Countermeasure development has thus far been focused on the well-documented problem of orthostatic intolerance, but no new validated countermeasures exist as yet to prevent this important operational problem. If arrhythmias and/or atrophy are determined to be important barriers to the health and well-being of crew members, then countermeasures will have to be developed to counteract them.

Most pulmonary studies are focused on the important issue of decompression sickness. Two key projects in this area (the clinical trial of reduced prebreathe protocols and in-suit Doppler monitoring) are not supported through either the NRA or the NSBRI programs, but are supported through operational funds. Other efforts recommended in the *Strategy* report (aerosol deposition, Martian dust effects, respiratory muscle function) are not currently in progress.

The development of the countermeasure evaluation plan and integrated testing regimens addresses important areas outlined in the *Strategy* report.

BIBLIOGRAPHY

Blomqvist, C.G., L.D. Lane, S.J. Wright, G.M. Meny, B.D. Levine, J.C. Buckey, R.M. Peashock, P. Weatherall, J. Stray-Gundersen, F.A. Gaffney, D.E. Watenpaugh, Ph.R.M. Arbeille, and F. Baisch. 1995. Cardiovascular regulation at microgravity. Pp. 688-690 in Proceedings of the Nordeney Symposium on Scientific Results of the German D-2 Spacelab Mission. P.R. Sahm, M.H. Keller, and B. Schieve, eds. Bonn and Köln, Germany: WPF and Deutsche Agentur für Raumfahrtangelegenheiten (DARA) and Deutsche Forschungsanstalt für Luft- und Raumfahrt.

National Aeronautics and Space Administration (NASA). 1997. Task Force Report on Countermeasures: Final Report. Washington, D.C.: NASA.

NASA. 1998a. Astronaut Medical Evaluation Requirements Document (AMERD), JSC 24834, Rev. A. Houston, Tex.: NASA.

NASA. 1998b. Life Sciences Program Tasks and Bibliography for FY 1998. Washington, D.C.: NASA.

NASA. 1999a. Critical Path Research Plan Presentation (including EDOMP results presented at that time). Committee on Space Biology and Medicine meeting, March 3-5, 1999. Houston, Tex.: NASA.

NASA. 1999b. Countermeasure Evaluation and Validation Project Plan. June 16, 1999. Houston, Tex.: NASA.

NASA and Universities Space Research Association (USRA). 1999. Proceedings of the First Biennial Biomedical Investigators' Workshop, January 11-13, 1999, League City, Texas. Houston, Tex.: NASA and USRA.

National Research Council (NRC), Space Studies Board. 1998. A Strategy for Research in Space Biology and Medicine in the New Century. Washington, D.C.: National Academy Press.

National Space Biomedical Research Institute (NSBRI). 1999. Proceedings of the Artificial Gravity Workshop, January 14-15, 1999, League City, Texas. Houston, Tex.: NSBRI.

6

Endocrinology and Nutrition

INTRODUCTION

Endocrinology is concerned with signaling between cells and tissues. Together with the nervous and immune systems, the endocrine system regulates the human response to spaceflight and the readjustment processes that follow landing. Quite apart from the intrinsic scientific interest of the underlying mechanisms, understanding the effects of spaceflight on the endocrine system is essential for the rational development of countermeasures.

The principal spaceflight responses that have a significant endocrine involvement are fluid shifts; perturbation of circadian rhythms; losses of red cell mass, bone, and muscle; and maintenance of energy balance. Endocrine and nutritional imbalances may also contribute to immune dysfunction, altered pharmacological responses, behavioral changes, and decreased resistance to radiation.

The *Strategy* report (NRC, 1998) identified three areas in endocrinology for investigation. The highest priority was assigned to obtaining a baseline in-flight human hormone profile. Lower-priority topics were refining and developing ground-based models and continuing the construction of a life sciences database that is accessible to the outside community. In the area of nutrition, the report focused on energy and protein requirements, with emphasis on the importance of maintaining energy balance in astronauts during spaceflight.

NASA'S CURRENT RESEARCH PROGRAM IN ENDOCRINOLOGY AND NUTRITION

This analysis is based primarily on the FY 1999 NASA Research Announcement (NRA) (98-HEDS-02), although any major program changes due to the approved projects resulting from the 1999 NRA solicitation (99-HEDS-03, funding starting October 1, 2000) have been identified. Also included in the analysis are projects from the National Space Biomedical Research Institute (NSBRI) that were active during FY 1999. Only three projects can be considered to be purely endocrine system research, although approximately sixteen other projects have significant endocrine components. This was about the same level of activity as the previous year. Two projects with significant endocrine involvement were approved from the 1999 NRA solicitation. To estimate the expenditure on endocrine research, the fraction of a project that is endocrine related has been multiplied by the total dollar amount of the grant.

The total expenditure in FY 1999 on endocrine-related themes was about $1.225 million (~3.4 percent of total NASA expenditures). About half ($650,000) of the funds were used to support studies of the endocrine aspects of circadian rhythm dysfunction (three projects). The total expenditure in FY 1999 was unchanged from 1998, although there was a decrease in the number of projects funded (21 versus 16). Table 6.1 summarizes the funding and project distributions between subdisciplines and between the NRA and NSBRI programs for FY 1999.

The *Strategy* report (NRC, 1998) identified circadian rhythm dysfunction as being a high-priority area for investigation. Two circadian rhythm studies were included as part of the Bion program in 1998 but were discontinued in FY 1999 ($100,000). The current program encompasses three ground-based human studies. The total expenditure on the endocrine aspects of circadian rhythm and sleep studies ($650,000) was unchanged from FY 1998. (For additional discussion of circadian rhythms and the relevance of their study to NASA, see Chapter 9.)

Expenditure on the endocrine component of other programs with endocrine involvement was about $575,000 in FY 1999 (13 projects). No currently funded projects relate to the highest-priority objective, obtaining a baseline in-flight human hormone profile. The Integrated Testing Regimen (ITR) does not meet this requirement. The lack of activity is a consequence of the current lull in flight opportunities. The second objective, the refinement and development of ground-based models, is more nebulous because model development is not a specifically stated objective of any project. For the most part, endocrine measurements are secondary objectives. The endocrine components of currently funded studies are directed toward elucidating underlying mechanisms, as recommended in the *Strategy* report. The greatest number of interdisciplinary projects with an endocrine component in FY 1999 were in the areas of bone (five studies), muscle (five studies), sleep and circadian rhythm (three studies), and

TABLE 6.1 Summary of FY 1999 Funding for Subdisciplines for Endocrinology, Nutrition, and Related Fields

Subdiscipline	NRA		NSBRI	
	Total ($ thousands)	No. of Projects	Total ($ thousands)	No. of Projects
Endocrinology	1,225	9	600	5
Sleep	650	3	0	0
Bone	300	3	150	2
Muscle	25	1	450	3
Cardivascular	50	2	0	0
Fluid/electrolyte	200	1	0	0
Nutrition	325	3	0	0
Gender	200	1	0	0
Thermoregulation	200	1	0	0
Pharmacology	150	2	0	0
Total	2,100	16	600	5

NOTE: For most of the projects listed under endocrinology, endocrinology is only a part of the project. To estimate the expenditure on the endocrine aspects, the total sum expended on the project has been multiplied by the estimated percentage of endocrine involvement.

cardiovascular (two studies). Where appropriate, investigators include endocrine measurements in their studies.

Although questions related to potential gender-specific effects of spaceflight have received some public interest, there is little research in this area. Two studies concerned with gender effects were carried out in FY 1998, a ground-based study of orthostatic intolerance at Johnson Space Center (JSC) and a student fellowship to study the response to exercise. In FY 1999, only the former was continued. There were also two ground-based studies on the cardiovascular aspects of thermoregulation in 1998 of which only one was continued into FY 1999.

The NRA and NSBRI programs for the two years for which detailed information is available involve little research in nutrition, and nutrition was the smallest of the research disciplines identified by NASA as a program area. Expenditure on nutrition was about $430,000 in 1998 and $325,000 in 1999. In 1998 there were five projects—vitamin D, assessment of body composition, gastrointestinal function, and two projects on renal stone formation. In 1999, two of the nutrition projects were discontinued. Related studies involve water balance (discontinued in 1999) and thermoregulation. However, there appears to be a trend in the FY 2000 NRA grants toward increasing activity in nutrition research in response to the recommendations in the *Strategy* report. Three projects have nutrition as their primary focus (two dealing with protein-energy balance and one with vitamin E), and three other projects have significant nutrition involvement. The JSC metabolism section has a clinical program to provide astronauts with balanced and acceptable diets. The program does not seem to have research components, although members of this group are involved in NASA- or NSBRI-funded research in other disciplines.

The endocrine component of the current NRA and NSBRI research programs is small. This is consistent with the *Strategy* report conclusions and appropriate given the lack of flight opportunities. Absence of flight opportunities precludes any work being done on the highest-priority objective, obtaining a baseline in-flight human hormone profile. The nutrition program remains small but, as recommended in the *Strategy* report, does appear to be receiving more attention.

PROGRAMMATIC BALANCE

Balance of Subdiscipline Areas

Given the current lack of flight opportunities, the small size of the direct investment in endocrinology at present is consistent with the *Strategy* report. The *Strategy* report identified endocrine measurements as important components of other disciplines (e.g., bone, muscle, circadian rhythms), and this is reflected in the distribution of endocrine measurements within the program. However, within the endocrinology discipline there may be an overemphasis on circadian rhythm studies (see Chapter 9). There is currently no activity related to acquiring a comprehensive hormonal profile of humans adapted to spaceflight, the principal recommendation of the *Strategy* report.

The issue of gender is important and does not appear to be addressed adequately. Prior studies have shown decisively that there are gender effects. A previous NASA-funded bed rest study showed that cortisol was elevated in males, but not females, during one week of bed rest (Vernikos et al., 1993). Except for the study on orthostatic hypotension, gender has not been a consideration in other studies where it might be relevant, for example, studies of bone loss.

Including projects approved from the 1999 NRA solicitation (funding to start in FY 2000), ongoing investigations include studies of calcium metabolism (one study), vitamins D (one study) and E (one study) metabolism, renal stone formation (two studies), body composition by bioimpedance (one study), gastrointestinal function (one study), and protein-energy balance (two studies). All of these projects address areas identified in the report as being of high priority; four astronauts are now known to have developed renal stones. It would appear that the recommendations of the *Strategy* report have been followed, the number of studies in nutrition has been increased, and the recent NRA (September 1999 for FY 2000) continues to list nutrition as a priority area.

Overall, the program is balanced and the distribution of projects between the various subdisciplines reflects the recommendations of the *Strategy* report. Specifically the trend for an increase in nutrition is consistent with the *Strategy* report. Baseline studies of the in-flight human endocrine profile will not be feasible until flight opportunities become available. Gender effects, although not a specific recommendation in the *Strategy* report, should receive more emphasis.

Balance of Ground and Flight Investigations

The small research program in endocrinology is predominantly ground based because of a lack of flight opportunities. Two of the three strategy recommendations require flight experiments, namely, determination of the in-flight human endocrine profile and further refinement of ground-based models.

The 1998 circadian rhythm projects were well balanced between flight- and ground-based studies and between human and nonhuman studies, principally because of Neurolab projects. The Neurolab projects were completed in FY 1998, so there was minimal activity in the flight program for sleep and circadian rhythms in FY 1999 although one project was approved for FY 2000. One particularly important project was the first double-blind, first-in-flight clinical trial of a countermeasure utilizing melatonin, a potential countermeasure to regulate circadian rhythm. The experiment was flown on the Neurolab mission and completed in FY 1998. A follow-up study was approved for funding in FY 2000. Astronaut nutrition has to be studied during spaceflight; so with the paucity of flight opportunities, the balance within the small program is reasonable. One of the four FY 2000 NRA awards with a significant nutrition component was a flight experiment.

Gastrointestinal function is related to nutrition. The two studies in this area are concerned primarily with pharmacokinetics, an important and underrepresented topic. Three additional studies in pharmacokinetics were approved for FY 2000, of which one was a flight experiment. This shift toward flight experiments is the results of some nutrition and pharmacological studies becoming feasible during the early phases of International Space Station (ISS) activity.

Future endocrine experiments are mostly ground based. However in accord with the high priority given to nutrition studies by the *Strategy* report, some of the funded nutrition projects are flight experiments and have been selected for consideration for early flight opportunities.

Emphasis Given to Fundamental Mechanisms

Endocrine studies play a supporting role in many areas. The ground-based work has the appropriate emphasis on mechanistic studies. Gone are the descriptive projects of uncontrolled phenomena and the random measurements of random plasma hormones from random samples. The nutrition studies recommended in the *Strategy* report are directed toward the maintenance of nutritional adequacy. **The foci of the endocrine and nutrition programs are appropriate and in concordance with the recommendations of the *Strategy* report.**

Utilization and Validation of Ground and Animal Models

The *Strategy* report repeatedly stressed the importance of validating and refining ground-based models. Current ground-based models can reproduce most of the symptoms of spaceflight, but there is not necessarily a commonality of mechanism. There is a critical unmet need for ongoing evaluation and refinement of existing models through comparison with flight data and endocrinological measurements, including a comprehensive in-flight human hormone profile.

When ground and flight studies have been compared, the results have sometimes been surprising. For example, a flight study on the Life and Microgravity Spacelab shuttle mission by Wronski et al. (1998) attempted to address the effects of glucocorticoids on rat bones during spaceflight, since there was ample ground-based data to suggest that glucocorticoids had a major role in bone loss. No effect of spaceflight or of cortisol on bone growth was found. Analysis of this unexpected finding showed that changes in bone in rats induced by spaceflight were influenced by (1) the way the animals were housed, (2) the age of the animal, (3) the particular strain (genetic background) of the rat used in the study, and (4) flight duration. Group housing inhibited bone formation, whereas single housing did not. Instead of being simple, the rat model turns out to be complex!

The majority of ongoing studies use ground-based models. Even though the *Strategy* report identified validation of ground-based models as a high-priority area, there is little activity aimed at validation or research to develop new models.

DEVELOPMENT AND VALIDATION OF COUNTERMEASURES

One of the bed rest countermeasure studies, the use of alendronate to reduce bone calcium loss, involves a series of measurements to evaluate any endocrine perturbations induced by the drug. The study involves a close interaction between university investigators (NRA funded), the NSBRI, and the JSC countermeasures task group. Endocrine measurements are also minor components of two other studies, which address the efficacy of resistive exercise in reducing bed rest-induced muscle atrophy. All of these studies have been well planned and are being done in general clinical research centers, where diet is controlled and the number of subjects is adequate for statistical analysis.

Apart from the validation of ground-based models, some countermeasures, particularly those involving drugs or hormonal manipulation, might have some long-term effects on other systems. For example, a potential scenario of secondary impact is the possibility of female astronauts being placed on oral contraceptives for regulation of the menstrual cycle. This may lower estrogen levels and promote bone loss in an already vulnerable population. There does not appear to be any consideration of such side effects in either the 1998 or the 1999 program, although it does appear to have been identified as a high-priority area for investigation in the FY 2000 NRA solicitation. Investigation of treatment interactions should be part of every countermeasure study.

In the area of nutrition, four studies are targeted at countermeasures. Two relate to countermeasures against in-flight renal stone formation with the objective of developing a dietary intake (fluid and electrolytes) and urine electrolyte excretion profile to predict the probability of renal stone formation. A third study is focused on vitamin E as an antioxidant, and the fourth on protein-energy balance. All of these studies are appropriate. There is a need to minimize the risk of renal stone formation. Oxidative damage has been implicated as a factor in chronic ground-based disease states such as cancer and atherosclerosis, so a study of the use of vitamin E as an antioxidant is relevant. The energy balance study is targeted at elucidating the mechanisms of the inability to maintain energy balance in flight.

Endocrine involvement in the countermeasures program is mostly consistent with the *Strategy* report. Two areas that require further emphasis are treatment interactions and gender effects.

The nutrition program is focused almost exclusively on countermeasures, and this is in concordance with the *Strategy* report.

EPIDEMIOLOGY AND MONITORING

The *Strategy* report assigned a high priority to obtaining an in-flight hormone profile. The database on the astronaut corps collected and maintained by the Space Medicine Program at JSC contains a substantial amount of endocrine data. The NASA list does not appear to match the likely requirements for the in-flight hormone profile. Having an external panel of experts would be helpful in deciding which hormones should be measured. **NASA should ensure that the database being compiled by medical operations at JSC is congruent with the *Strategy* report recommendation for determining the in-flight human hormone profile.**

SUPPORT OF ADVANCED TECHNOLOGIES

For the most part, new methodologies—particularly sensitive micromethods—are not being used in individual research programs. At the First Biennial Life Sciences Investigators' Workshop, NASA (Ames) presented an exhibit of some advanced technologies, including the use of miniature time-of-flight mass spectrometers. **At present, the endocrine programs do not appear to be taking full advantage of the advanced technologies that have been developed to make endocrine measurements for medical studies on patients.**

SUMMARY

Given the current lack of flight opportunities, the small endocrine program is reasonable. Two of the three recommendations in the *Strategy* report require flight data that currently cannot be obtained. The *Strategy* report identified determination of the in-flight endocrine profile as the highest priority for endocrine research. In the absence of flight opportunities, this goal is at present unattainable. However, no plan has been developed for obtaining a set of core data on the human endocrine response to living in low Earth orbit. Knowledge and understanding of endocrine changes with flight are essential for the refinement of ground-based models. In other studies, endocrine measurements are being made where appropriate.

Although in the past the nutrition program was inadequate, during the last two years NASA,[1] the NSBRI, and this committee (*Strategy* report) have independently identified nutrition-energy balance as a very high priority area for future studies. NASA appears to have responded to the *Strategy* report by giving increased emphasis to research in nutrition. Inadequate attention is being paid to gender issues and treatment interactions.

REFERENCES

National Research Council (NRC), Space Studies Board. 1998. A Strategy for Research in Space Biology and Medicine in the New Century. Washington, D.C.: National Academy Press.

Vernikos, J., M.F. Dallman, L.C. Keil, D. O'Hara, and V.A. Convertino. 1993. Gender differences in endocrine responses to posture and 7 days of 6-degrees head-down bed rest. Am. J. Physiol. (Endo. and Metab.) 265:E153-61.

Wronski, T.J., M. Li, Y. Shen, S.C. Miller, B.M. Bowman, P. Kostenuik, and B.P. Halloran. 1998. Lack of effect of spaceflight on bone mass and bone formation in group-housed rats. J. Appl. Physiol. 85:279-85.

---

[1] As indicated in NASA's Critical Path Research Plan presentation (including EDOMP results presented at that time) at the Committee on Space Biology and Medicine meeting, March 3-5, 1999, Houston, Texas.

7

Immunology and Microbiology

INTRODUCTION

The immune system plays a fundamental role in protecting the host against the development and persistence of infections and tumors. There have been a wide range of alterations in measures of cell-mediated immune function after long-term and short-term spaceflight (alterations in cytokine production, inhibited leukocyte proliferation, inhibited skin test reactions that measure immune memory, alterations in distribution of leukocytes, inhibited natural killer cell activity, and decreased responses of bone marrow cells to colony-stimulating factors) (Sonnenfeld, 1998). These are standard measures of immune function. Although there can be little doubt after 20 years of work that immune responses are altered after exposure to spaceflight conditions, the physiological and medical significance of these spaceflight-induced changes for crew health remains unknown. There have been only limited reports of difficulty with infections during or after spaceflight, and infectious diseases have not been a major issue. Nevertheless, with the advent of longer-term spaceflights, including a permanent presence on the International Space Station (ISS) and other longer flights, the potential for development of infectious diseases in crews may increase as a result of prolonged alterations in immune parameters. Changes in the composition of indigenous microbial flora during long-term spaceflight and, perhaps, in the properties of microorganisms and the effectiveness of antimicrobial drugs also could potentially pose problems.

Immunology and microbiology are linked together because of the fundamental role of the immune system in resistance to infection. However, alterations in immune responses could also lead to changes in tumor immunity and to the development of allergic and autoimmune diseases, either in flight or postflight.

Changes in the immune system and in the properties of microorganisms induced by spaceflight conditions have been recognized by NASA and others as an area for further study (NASA, 1998; NRC, 1998). In the 1998 *Strategy* report (NRC, 1998), major priorities included (1) determining whether alterations in immune responses induced by spaceflight conditions actually resulted in compromised resistance to infection and (2) defining the interactions of the immune systems with other systems, in particular the neuroendocrine system, that could contribute to compromised resistance to infections and tumors in spaceflight. These recommendations are discussed in greater detail in Appendix A.

NASA'S CURRENT RESEARCH PROGRAM IN IMMUNOLOGY AND MICROBIOLOGY

In FY 1999, nine grants for immunology and microbiology (including funding for the National Space Biomedical Research Institute (NSBRI)) were funded, at a level of approximately $1.160 million (see Table 7.1). These include investigations at Johnson Space Center (JSC), the NSBRI program, and other extramural NASA Research Announcement (NRA)-supported laboratories. Ames Research Center (ARC) is not currently involved in studies in immunology and microbiology. Late in 1999, four new immunology-related projects were approved through the Gravitational Biology and Ecology (GB&E) program of the NRA process for a total funding of about $265,000. Studies are well coordinated between JSC and NSBRI laboratories and between JSC and independent university laboratories for the immunology and microbiology disciplines.

For FY1999, one additional grant was funded by the Biomedical Research and Countermeasures (BR&C) program, and five basic research grants were funded by the GB&E program. There is no specific request for proposals in immunology or microbiology in the FY 2000 NASA Research Announcements for Biomedical Research and Countermeasures, but the call for physiology experiments could encompass immunology and microbiology (NASA, 1999b). Immunology and microbiology remain areas of relatively low priority for NASA-funded research.

There is considerable overlap and integration between the fields of immunology and microbiology. Experiments focus on the effects of spaceflight on immunological parameters. In-flight and ground-based studies in analogue settings, such as Antarctic overwintering, on human immune responses and resistance to infections are ongoing and planned. Additionally, an animal model (rodent hindlimb unloading) is being used to test the effects of spaceflight on immune responses and resistance to viral and bacterial infection. There are also studies using animal models to determine possible effects of spaceflight on the development of immune responses. In addition, there are very active ongoing research programs on the rapid identification of microorganisms and the effects of the spaceflight environment and ground-based models on microbial growth that are supported by NSBRI and NASA.

The priorities outlined in the *Strategy* report are beginning to be considered. Studies to determine whether changes in immune parameters induced by spaceflight conditions affect resistance to infection were a priority of the *Strategy* report. JSC, NSBRI, and NRA-funded university investigators have undertaken both ground-based and spaceflight studies to address the question of reactivation of latent

TABLE 7.1 Summary of FY 1999 Funding for Immunology and Microbiology

Subdiscipline	NRA		NSBRI	
	Total ($ thousands)	No. of Projects	Total ($ thousands)	No. of Projects
Immunology	135	2	365	1
Microbiology	330	4	330	2
Total	465	6	695	3

NOTE: The bulk of NRA funding for immunology and microbiology comes from the GB&E program (now the FBRP) and not from the Biomedical Research and Countermeasures program.

viral infections. These should begin to provide answers as to whether spaceflight-induced changes in immune parameters can affect resistance to infection. This approach is in very preliminary stages and should be expanded. The FY 1999 NRA awards include a flight study to explore this issue.

Interdisciplinary investigation of the interaction of the neuroendocrine system hypothalamus-pituitary-adrenal (HPA) axis with the immune system under spaceflight conditions was a second priority of the *Strategy* report. Although some studies have been carried out in this area, there is no directed program to address this issue.

The *Strategy* report did not specifically address priorities for microbiology research. However, the approach to develop new techniques for detection of microbes and new approaches for disinfection is appropriate to empower future space exploration.

In summary, the current activities in immunology and microbiology are congruent with the recommendations of the *Strategy* report, except that minimal effort is directed at the interaction of the immune system with the HPA axis. Immunology and microbiology remain areas of relatively low priority for NASA-funded research.

PROGRAMMATIC BALANCE

Balance of Subdiscipline Areas

The *Proceedings of the First Biennial Space Biomedical Investigators' Workshop* (NASA and USRA, 1999), the NASA Life Sciences Program Tasks (NASA, 1999a), and NRA awards for FY 1999 (NASA, 1999b) indicate an appropriate balance between immunology and microbiology. For FY 1999, there were three funded immunology grants (total funding approximately $500,000) and six microbiology tasks (total funding $660,000). There is overlap between the subdisciplines, and several microbiology projects contain immunological components. The major focus of the immunology and microbiology program is investigation of whether changes in immune responses in spaceflight result in alterations in resistance to infection. An additional focus of the program is rapid detection of microorganisms.

Balance of Ground and Flight Studies

The majority of investigations are ground based (NASA and USRA, 1999), due in part to reduced opportunities for spaceflight as a result of ISS construction. There are currently two experiments scheduled for spaceflight, one involving cell culture experiments on the effects of spaceflight on macrophage maturation (NASA, 1999a), the other aimed at determining the effects of spaceflight on virus reactivation (NASA headquarters, personal communication, 1999). The date for flight of these experiments is not yet available. NSBRI-funded research in immunology and microbiology is currently limited to ground-based studies, in accord with the institute's charter (NSBRI, 1998).

Emphasis Given to Fundamental Mechanisms

The NSBRI group and NASA-funded investigators have begun studies on fundamental mechanisms involved in spaceflight-induced alterations in immune responses, using cell culture, human, and rodent models. These include studies on the mechanisms of immune cell function and the mechanisms of the effects of rodent hindlimb unloading on delayed hypersensitivity. Additional studies to determine new methods for the detection of bacteria in extreme environments are also of a fundamental nature. The study of these fundamental mechanisms is in accord with the recommendations of the *Strategy* report.

There appears to be little attention currently given to the role of the HPA axis in spaceflight-induced changes in resistance to infection. Studies analyzing interactions between the HPA axis and the immune system in modulating resistance to infection are not currently being carried out, although such studies were accorded high priority in the *Strategy* report.

Utilization and Validation of Animal Models

There has been some validation of animal models. Spaceflight-induced changes in immunological parameters are similar in humans and rodents (Sonnenfeld, 1998). Ground-based experiments with the rat hindlimb-unloading model and simultaneous spaceflight experiments showed generally similar immunological changes, except for leukocyte subset distribution (Sonnenfeld et al., 1992). Rodent hindlimb unloading is currently being used by NSBRI investigators to examine resistance to viral infections. There appear to be no plans as yet for flight experiments using rodent or other animal model systems (NASA, 1998, 1999b).

In summary, there is appropriate balance between the immunology and microbiology subdisciplines, and appropriate emphasis on the study of fundamental mechanisms. Use and validation of animal models are being carried out as outlined in the *Strategy* report. The balance between ground-based and flight studies is congruent with the recommendations of the *Strategy* report to carry out precursor ground-based studies. Determination of the physiological and biomedical significance of spaceflight-induced changes in immune responses should include future flight experiments.

DEVELOPMENT AND VALIDATION OF COUNTERMEASURES

There is an effective countermeasure available for potential problems with respiratory tract infections. The Health Stabilization Program developed as part of the Apollo Program has diminished the frequency of respiratory tract infections in crews (Hawkins and Ziegschmid, 1975), probably due to limitations in preflight social contacts. As the duration of spaceflight increases, the risk of infection may also increase if immune function is compromised. The goal of the NSBRI is to develop additional countermeasures to deal with potential problems of infection that may be encountered during and after long-term spaceflight (NSBRI, 1998). **The existing countermeasure for respiratory tract infections has proven effective. Development of further countermeasures should await confirmation of the physiological and biomedical significance of spaceflight-induced changes in immune responses, as recommended in the *Strategy* report.**

EPIDEMIOLOGY AND MONITORING

Monitoring of respiratory tract infections during the Apollo Program led to the Health Stabilization Program for Astronauts, which has decreased the incidence of infections (Hawkins and Ziegschmid, 1975). The only ongoing monitoring and epidemiology studies are those involving the crew health program (NASA, 1998), for which data have not been generally available. The proposed Integrated Testing Regimen (ITR) for spaceflight crews provides for microbiological (parameters established) and immunological (parameters to be determined) monitoring. **If information from the Integrated Testing Regimen is made available to the research community and the parameters chosen for immunological monitoring are appropriate, then there may be congruence with the *Strategy* report recommendation for development of a monitoring program. There does not appear to be a program for epidemiological studies available for research purposes.**

SUPPORT OF ADVANCED TECHNOLOGIES

Rapid detection of microorganisms is of prime importance in ensuring astronaut health for the entire space program. To meet this need, NASA, in collaboration with industry, is supporting the development of several new technologies for the detection of microorganisms during spaceflight (NASA,1999a; NASA and USRA, 1999; NSBRI, 1998). **The support for advanced technology development for microbiology is congruent with the general recommendation of the *Strategy* report. There is no current technology development specific to immunology.**

SUMMARY

The immunology and microbiology program has begun to develop priorities in accordance with those suggested in the *Strategy* report for ground-based experiments using animal models prior to the development of spaceflight studies. The extensive use of models by the programs, including animal models such as hindlimb unloading of rodents, tissue culture in the rotating cell vessel apparatus, and analogue environments such as isolation and polar overwintering, should permit critical testing of mechanistic hypotheses in space. Additional efforts should be directed toward such mechanistic studies and toward studies involving interactions with the HPA axis.

The number of investigators and funded projects for the entire immunology and microbiology programs is small, and the need for additional studies will become more pressing as prolonged exposure of crews to spaceflight conditions becomes the norm.

REFERENCES

Hawkins, W.R., and J.F. Ziegschmid. 1975. Aspects of crew health. P. 43 in Biomedical Results of Apollo. R.F. Johnston, L.F. Dietlen, and C.A. Berr, eds. SP-368. Houston, Tex.: NASA.

National Aeronautics and Space Administration (NASA), Space and Life Sciences Directorate. 1998. Astronaut Medical Evaluation Requirements Document–Appendix B, Selection, Annual, Pre- and Postflight Medical Evaluation Requirements. Houston, Tex.: NASA.

NASA, Office of Life and Microgravity Sciences and Applications. 1999a. Life Sciences Program Tasks and Bibliography for FY 1998. Washington, D.C.: NASA.

NASA. 1999b. Research Announcement—Research Opportunities in Space Life Sciences. Biomedical Research and Countermeasures. Washington, D.C.: NASA.

NASA and Universities Space Research Association (USRA). 1999. Proceedings of the First Biennial Space Biomedical Investigators' Workshop, January 11-13, 1999, League City, Texas. Houston, Tex.: NASA and USRA.

National Research Council (NRC), Space Studies Board. 1998. A Strategy for Research in Space Biology and Medicine in the New Century. Washington, D.C.: National Academy Press.

National Space Biomedical Research Institute (NSBRI). 1998. Annual Report, October 1, 1997-September 30, 1998. Houston, Tex.: NSBRI.

Sonnenfeld, G. 1998. Immune responses in space flight. Int. J. Sports Med. 19:S195-S2.

Sonnenfeld, G., A.D. Mandel, I.V. Konstantinova, W.D. Berry, G.R. Taylor, A.T. Lesnyak, B.B. Fuchs, and A.L. Rakhmilevich. 1992. Spaceflight alters immune cell function and distribution. J. Appl. Physiol. 73:191S-195S.

8

Radiation Biology

INTRODUCTION

Concerns about the acute and chronic health effects of the radiation encountered in space (low Earth orbits and the International Space Station (ISS), and in extraterrestrial space) have been raised for many years. The risks arise from the interactions between high-energy charged particles, ranging from protons to iron nuclei, with the DNA, cells, and tissues of humans. Such interactions may kill cells; cause them to mutate; induce cancers; and injure the central nervous system (CNS), the immune system, and the reproductive system. Predicting the probability of such damaging effects requires a knowledge of dosimetry and radiation biology. Dosimetry is concerned with the type, energy, and number of particles in space; how their distribution changes in passing through spacecraft shielding; and how they depend on spacecraft location and time. Location in low Earth orbit (LEO) strongly affects the spatial distribution of the particles, and large solar particle events (SPEs) may increase the flux of protons by several orders of magnitude for many days (Badhwar, 1997). SPEs are associated with very strong increases in the Sun's magnetic field, which result in highly significant decreases in the galactic cosmic-ray background. The SPEs are much more frequent during the period of solar maximum; hence the flux of galactic cosmic rays is larger during solar minimum. Radiation biology research deals with the effects of solar and cosmic-ray particles on suitable model biological systems that may be used to extrapolate risks to humans.

The 1998 *Strategy* report (NRC, 1998) gives estimates of the uncertainty in the carcinogenic risks from high-atomic-number, high-energy (HZE) particles that are between a factor of 4 and 15. The reason for the large uncertainty is that there is only one ground-based carcinogenesis experiment, on cancer induction in mice, that estimates the risk from HZE particles relative to gamma rays as appreciably greater than the risks determined for cell killing, mutation, or transformation of cells exposed in vitro (Alpen et al., 1993). Hence quantitative designs of appropriate countermeasures, such as shielding, and biochemical or biological schemes to reduce the damage from HZE particles or augment repair following radiation exposure, are very rudimentary. The *Strategy* report recommended a comprehensive research program to determine the risks from different types and energies of HZE particles and from high-energy protons for a number of biological end points and the efficacy of different types and thicknesses of shielding in reducing these risks. These recommendations are described in more detail in Appendix A.

NASA'S CURRENT RESEARCH PROGRAM IN RADIATION BIOLOGY

NASA's *Strategic Program Plan for Space Radiation Health Research* (NASA, 1998) was approved in October 1998 by the associate administrator for the Office of Life and Microgravity Sciences and Applications. The budget breakdown for FY 1999 is given in Table 8.1. The program follows closely the principal recommendations of the *Strategy* report, which were given as experimental procedures designed to answer higher- and lower-priority research questions. The higher-priority recommendations were aimed at determining the carcinogenic risk and effects on the CNS of exposure to energetic protons and HZE particles; how the composition of the shielding would quantitatively ameliorate the biological effect of HZE particles; and whether there are studies on radiation-induced genetic changes that could increase confidence in extrapolating from rodents to humans and might enhance a similar extrapolation for cancer. Other high-priority recommendations were to determine if there were better analyses that could decrease the present uncertainties in the risks of HZE effects, to determine how the design of the space vehicle could affect the internal radiation levels, and to determine whether SPEs could be predicted with sufficient advance warning for astronauts to return to a shielded shelter. Low-priority recommendations were to estimate the effects of long-duration flight on fertility and on cataract formation, to

TABLE 8.1 Summary of FY 1999 Funding for Radiation Research by Subdiscipline

Subdiscipline	NRA Total ($)	NRA No. of Projects	NSBRI Total ($)	NSBRI No. of Projects	NSCORT Total ($)	NSCORT No. of Projects
Dosimetry						
Instrumentation	460,000	2				
Energy loss or scattering	75,000	1				
Russian plutonium workers	150,000	—				
Biological effects						
Carcinogenesis	200,000	1	523,819	3		
Cataracts	297,000	1				
Cell cycle	223,962	1				
Mutagenesis	234,000	1				
Cytogenetics	171,350	1	265,616	2		
Genomic instability[a]	1,000,000	1				
Radiation research	77,000	1			1,045,378	3
Total	2,888,312		789,435		1,045,378	

NOTE: In addition to the values given in the table, the funding for radiation studies includes $3.5 million to begin construction of the BAF, $1.075 million to support the operation of the present HZE operations at the AGS, and $6.5 million of directed spending for operation of the proton facility (< 250 MeV) at Loma Linda University.

[a]$1 million for cooperative research with the National Cancer Institute.

determine whether drugs could be used to protect against the carcinogenic effects of exposure to HZE particles, and to determine whether there are assays that might identify individuals with a predisposition for susceptibility to cancer, and whether biological responses to HZE particles depend only on the linear energy transfer (LET) or on the values of the atomic number and energy separately. Carrying out these experiments on molecules, on living cells in vitro and in vivo, and on animals so as to extrapolate to human risks requires ready access to energetic beams (up to several GeV per nucleon) of a number of nuclei that could mimic galactic cosmic rays. The major facility for these experiments is the Alternating Gradient Synchrotron (AGS) at Brookhaven National Laboratory, but it is available for only two to four weeks per year. At the present rate of progress it would take 20 or more years to complete the high-priority experiments recommended in the *Strategy* report, because of the limited accelerator time to carry out HZE particle and energetic proton irradiations at 1 GeV per nucleon. The *Strategy* report recommended that other sources of HZE particles be identified or constructed. NASA has followed this recommendation with the start of construction of a new accelerator beam line, the Booster Application Facility (BAF), to be built at Brookhaven National Laboratory with NASA funds. This facility will make available, for a significant fraction of the year, research time for HZE experiments on biological systems. Other useful accelerators in Japan and Germany will be recruited for similar efforts but at lower energies and atomic numbers.

In future years, the essential increase in funding for construction of the BAF could seriously compromise the rest of the biomedical research program, unless additional monies are made available. NASA should make every effort to support appropriately both the research and the facilities to do this research.

NASA's Strategic Program Plan (NASA, 1998) is very clear in pointing out that "current knowledge of radiation effects in space is not adequate for the design of long-duration missions without incurring either unacceptable risks or excessive costs."

In 1998 and 1999, the total program in radiation biology and dosimetry included 35 NASA Research Announcements (NRAs), 3 NASA Specialized Centers of Research and Training (NSCORT), and 5 National Space Biomedical Research Institute (NSBRI) projects or subprojects. The First Biennial Space Biomedical Investigators' Workshop (January 1999) included descriptions (abstracts) of 23 relevant projects, of which only three dealt with research on vertebrates—one on mutations of an exogenously incorporated gene in mice; one on behavioral effects of HZE particles on rats; and one on the induction of breast cancer in rats by HZE particles and gamma rays. The other projects dealt with dosimetry (seven projects); HZE effects on molecules, chromosomes, and cells in vitro (ten projects); and HZE effects on molecules, chromosomes, and cells in vivo (one project) (see NASA and USRA, 1999). There was a similar distribution in the *Abstracts of the 10th Annual Space Radiation Health Investigators' Workshop* (June 13-16, 1999) (20 on dosimetry, 25 on cells in vitro, 5 on cells in vivo, and 3 on animals). This distribution of projects is not that envisaged in the *Strategy* report (see below).

In addition, NASA holds annual 4-day workshops that bring together investigators on radiation effects to discuss the latest results in sessions of invited and proffered papers.

NASA has made an excellent start in following the recommendations in the *Strategy* report and in starting construction of a dedicated beam line that will operate for a large fraction of the year to supply HZE particles for biological experiments.

PROGRAMMATIC BALANCE

Balance of Subdiscipline Areas

There is adequate balance in the radiation program between dosimetry and biology. Extensive work is being done at JSC on real-time dosimetry in LEO, external and internal to a spacecraft. This work also supports a recommendation made in a recent NRC report (NRC, 2000), which also recommends implementaion, by the space physics community, of steps to predict SPEs and their intensities and locations using real-time data. Retrospective personal dosimetry is being carried out using measurements of chromosome translocations in lymphocytes of astronauts following spaceflight, compared to a baseline of aberrations before flight (Yang et al., 1997). Chromosome translocations in peripheral blood lymphocytes are excellent estimates of cumulative exposure. However, translocations as a function of dose depend on both the dose rate and the nature of the incident charged particles—their charges, energies, and linear energy transfers in tissue (Straume and Bender, 1997). Thus, for chromosome dosimetry that will give the dose equivalent of the radiation field in the space environment, one must know the nature of the particles as a function of time in flight and the translocation yields for these particles, measured on Earth at high dose rates and extrapolated to the dose rates observed in space. Data to be used for extrapolating from high dose rates to low dose rates cannot be obtained by exposing lymphocytes in vitro because of the short lifespan of these cells in vitro. The data will have to be obtained by exposing lymphocytes in vivo using a suitable animal model, probably monkeys.

At present there is no provision for extrapolating from the acute experimental calibrations on Earth to the chronic exposures usually experienced in space.

Experiments at ground-based accelerators are being carried out to determine the effectiveness of various types and thicknesses of shielding in reducing doses from HZE particles, for various biological end points.

The written plan (NASA, 1998) is similar to the *Strategy* report, but there is a marked difference in emphasis. The report emphasizes carcinogenic end points and molecular, chromosomal, and cellular effects from in vivo exposures to HZE particles and high-energy protons. Only one experiment is being carried out with a relevant cancer end point—breast cancer induction in rats (the report recommends mice as the more appropriate species). Although the biological research uses state-of-the-art techniques in molecular and cellular effects, it is not clear how data from these experiments will be used to estimate the relative biological effectiveness (RBE) of HZE particles for cancer induction. These molecular and cell biology experiments were proposed in the *Strategy* report as a way to validate the general biological effects of radiation fields behind shielding, but no such shielding experiments have yet been carried out.

Only one project deals with the effect of HZE particles on the CNS, in terms of changes in the behavior of rats following exposure. This type of research was a high priority in the *Strategy* report, and it was estimated that such a project could take 5 to 7 years if 3 months of beam time were available per year. This is probably an underestimate given that a recent study, not related to radiation effects, that involved mouse behavior and was carried out in three separate laboratories with genetically identical mice and the same experimental protocols, gave three different results (Crabbe et al., 1999). Hence, NASA should increase its efforts in CNS effects and in the use of state-of-the-art molecular and brain-scanning techniques, such as positron emission tomography, to measure the effect of energetic nuclei on brain chemistry.

The main and overriding issue to be solved before a spacecraft is designed is the appropriate shielding necessary to minimize the effects of exposure to radiation. The shielding design depends on HZE particle effects on the high-priority end points of carcinogenesis and CNS effects. Low-priority effects such as synergism with other variables and identification of in-flight markers of exposure (rather than determinations made on the ground) should not receive any emphasis.

Radiation limits to satisfy the principle of ALARA (as low as reasonably achievable) for LEO are currently being revised by the National Council on Radiation Protection and Measurements. The anticipated new career limits will be significantly lower than the current ones (NASA, 1998, p. 41). No such limits are currently available for travel beyond LEO. The results of the HZE experiments on molecules, cells in vitro, cells in vivo, and animals must be synthesized so as to predict the extrapolated risks to astronauts, as emphasized in the *Strategy* report. Such a synthesis procedure should be put in place in order to recommend radiation limits for astronauts exposed to HZE particles. These recommendations, plus knowledge of the types and energies of HZE particles, should guide the design of shielding.

Balance of Ground and Flight Investigations

The overwhelming majority of radiobiological experiments are, and will be, carried out on the ground and, of necessity, will use animals. The flux rate of HZE particles in space is low, and although the particles, if not appropriately shielded, could over a long period of time produce deleterious effects in humans, the use of such radiation for radiobiological experiments on small vertebrates in space is impractical because it is not possible to transport an HZE accelerator into space. Some radiation experiments in space, using sparsely ionizing radiation, have been carried out on human lymphocytes, microorganisms, and the small roundworm *Caenorhabditis elegans*. The radiation sources used were radioisotopes emitting β-particles. None of these experiments have shown any significant synergistic or antisynergistic effect of radiation and microgravity. Similar experiments on mice on the ISS would require an x-ray or γ-ray source to irradiate animals at 1 g and at microgravity. Since there is no compelling theoretical reason to expect that hypogravity will affect radiation end points in vertebrates, the committee's view is the same as that in the *Strategy* report—such experiments "with all their logistical difficulties, will not be rewarding and may not be worth the effort" (p. 190).

Hence, it is not clear how one might validate ground-based risk prediction in space by ISS utilization as suggested in NASA's Strategic Program Plan (NASA, 1998, pp. 17, 18, 20). The biological dosimetry behind shielding may be validated by ground-based experiments (see "Development and Validation of Countermeasures" below). On the other hand, the validation of biological risk estimates, behind shielding, of HZE nuclei and energetic protons cannot be done in space but must be done at a ground-based facility.

Emphasis Given to Fundamental Mechanisms

The effects of HZE particles on humans cannot be measured directly for ethical and practical reasons. The effects must be extrapolated from the results of animal experiments using fundamental knowledge of the mechanisms of radiation damage and repair. NASA strongly supports basic experiments on the effects of HZE particles on molecules, chromosomes, cells, and tissues so as to derive the necessary extrapolation rules to estimate risks to humans from exposure to HZE particles, as recommended in the *Strategy* report.

Utilization and Validation of Animal Models

The presumed important effects of HZE particles are possible carcinogenic and CNS effects. Estimates of the magnitudes of these effects must come from extrapolations from animal experiments as outlined in the previous two subsections of this chapter and in the *Strategy* report. The principal species used are mice and rats, for which typical measurement end points are chromosomal changes, cell killing, cellular apoptosis, mutations, cancer induction, CNS responses, and brain metabolic abnormalities.

There is a good balance between dosimetry and molecular and cellular radiobiology. However, more emphasis should be placed on carcinogenesis and CNS end points, using animal models. Such experiments must be carried out in ground-based facilities so as to estimate the risks to astronauts of exposure to HZE particles and develop guidelines for limits on exposure to these particles.

DEVELOPMENT AND VALIDATION OF COUNTERMEASURES

The proven countermeasures are shielding and operational parameters (especially in LEO), such as orbit paths and warnings of SPEs. However, calculated shielding (composition and thickness) must be validated, as described in the high-priority research questions in the *Strategy* report, by measuring the distribution and fluxes of particles behind shielding and then calculating and measuring their effects on simple biological end points such as cell killing and chromosome aberrations. Presumably, because of the limited time now available at HZE accelerators, such validation experiments have not been attempted or scheduled.

Other possible countermeasures, which are currently impractical, are (a) identification of individuals with high radiation resistance; (b) the use of chemical radioprotectors; (c) genetic methods to enhance the repair of radiation damage; and (d) interventions following unexpected radiation exposures that might enhance repair or induce apoptosis of damaged cells. These countermeasures are among the low-priority research questions of the *Strategy* report. NASA in collaboration with the National Cancer Institute should stimulate research in these areas.

Biodosimetry, by measuring chromosome translocations in lymphocytes, gives only retrospective doses and is not a countermeasure.

Autologous bone marrow transplants have been used on Earth to counteract acute radiation exposure. The proposal, in the NASA Strategic Program Plan, to use this as a countermeasure (NASA, 1998, p. 24) is of questionable feasibility. Presumably, exposed individuals will have to return to Earth for the procedure.

Major emphasis should be given to determining the types and thicknesses of shielding necessary to reduce astronaut risks to acceptable levels.

EPIDEMIOLOGY AND MONITORING

Measurements of chromosome translocations and their translation into dose equivalents (in sieverts) should be required for all long-term space travelers as indicated in NASA's Countermeasure Evaluation and Validation Project Plan (NASA, 1999). Data on translocations and on the methods used to calculate dose equivalents from these translocations should be available to compare with the subsequent monitored health of astronauts, as emphasized in the discussion of human flight in the *Strategy* report.

SUPPORT OF ADVANCED TECHNOLOGIES

NASA is a strong supporter of advanced and innovative technologies in radiation dosimetry, biodosimetry, and molecular and cellular biology as applied to radiation effects, as indicated by

the NSBRI and NSCORT programs and in their competitive, peer-reviewed research projects, which are reported at annual space radiation health investigators' workshops.

SUMMARY

At the present rate of progress it would take 20 or more years to complete the high-priority experiments recommended in the *Strategy* report, because of the limited accelerator times available to carry out HZE particle and energetic proton irradiations at 1 GeV per nucleon. NASA understands this problem and, to address it, has allocated initial funds to begin construction of a dedicated accelerator facility (the Booster Application Facility (BAF)) that, when completed, will supply the necessary energetic particles for the following decade and longer. Arrangements have been made to use other lower-energy facilities (Loma Linda: 250-MeV protons; HIMAC in Japan: 0.4 GeV per nucleon of HZE nuclei). The high costs of building and operating facilities could seriously deplete the funds for fundamental research in the many relevant and important biological, biomedical, physiological, and behavioral areas associated with long-term spaceflight.

The projects in high-priority radiation research areas—carcinogenesis and CNS—are at present poorly represented in the area of biological end points as determined in animals. The majority of projects are related to determinations of changes in cells, following exposure to HZE nuclei, in parameters such as turning genes on or off, cell cycle alterations, production of chromosome aberrations, mutation, and transformation. These experiments are state of the art, but it is not clear how the results will translate into helping estimate the radiation risks to astronauts.

Radiation dosimetry and computations and measurements of the radiation fluxes behind various types and thicknesses of shielding are well carried out, but they have to be validated by ground-based experiments on simple in vitro systems. The risk to astronauts of exposure to galactic cosmic radiation, estimated from ground-based experiments on molecules, cells, and animals, cannot be validated experimentally but could be approached by using several independent methods to calculate risks as proposed in the *Strategy* report.

REFERENCES

Alpen, E.L., P. Powers-Risius, S.B. Curtis, and R. DeGuzman. 1993. Tumorigenic potential of high-Z, high-LET charged-particle radiations. Radiat. Res. 136:382-391.
Badhwar, G.D. 1997. The radiation environment in low-Earth-orbit. Radiat. Res. 148:S3-S10.
Crabbe, J.C., D. Wahlsten, and B.C. Dudek. 1999. Genetics of mouse behavior: Interactions with laboratory environment. Science 284:1670-1672.
National Aeronautics and Space Administration (NASA). 1998. Strategic Program Plan for Space Radiation Health Research. Life Sciences Division, Office of Life and Microgravity Sciences and Applications. Washington, D.C.: NASA.
NASA. 1999. Countermeasure Evaluation and Validation Project Plan. Houston, Tex.: Johnson Space Center.
NASA and Universities Space Research Association (USRA). 1999. Proceedings of the First Biennial Space Biomedical Investigators' Workshop, January 11-13, 1999, League City, Texas. Houston, Tex.: NASA and USRA.
National Research Council (NRC), Space Studies Board. 1998. A Strategy for Research in Space Biology and Medicine in the New Century. Washington, D.C.: National Academy Press.
National Research Council (NRC), Space Studies Board. 2000. Radiation and the International Space Station: Recommendations to Reduce Risk. Washington, D.C.: National Academy Press.
Straume, T., and M.A. Bender. 1997. Issues in cytogenetic biological dosimetry: Emphasis on radiation environments in space. Radiat. Res. 148:S60-S70.
Yang, T.C., K. George, A.S. Johnson, M. Durante, and B.S. Federenko. 1997. Biodosimetry results from spaceflight Mir-18. Radiat. Res. 148:S17-S23.

9

Behavior and Performance

INTRODUCTION

The NASA Research Agenda for the International Space Station (NASA, 1998c) includes the following two questions: (1) how do microgravity and the space environment affect human behavior and performance, and (2) how can we enhance human performance in spaceflight? The inclusion of these questions in the ISS research agenda represents a substantial increase in emphasis on human behavior and performance in space among the biomedical research programs funded and supported by NASA. As noted in the *Strategy* report (NRC, 1998), studies conducted in ground-based analogue settings, as well as NASA's experience on the Shuttle Mir Space Program (SMSP), have provided ample evidence of significantly impaired psychosocial adaptation in space.

NASA's current efforts in the area of behavior and performance involve two different multi-disciplinary approaches: psychosocial-neurobehavioral and human factors engineering. Each approach is represented by one of two different administrative units within the Office of Life and Microgravity Sciences and Applications, with overlapping spheres of interest, authority, and expertise: Space Human Factors Engineering (SHFE) and Behavior and Performance (BP). Although both approaches are concerned with understanding factors that influence human performance during spaceflight and the development of countermeasures to enhance performance, this chapter addresses only the psychosocial-neurobehavioral approach.

The *Strategy* report identified two major priorities for research on psychological and social issues in long-duration spaceflight. First, research should be conducted on the neurobiological (circadian, endocrine) and psychosocial (individual, group, organizational) mechanisms underlying the effects of physical (microgravity, hazards) and psychosocial (isolation, confinement) environmental stressors on cognitive, affective, and psychophysiological measures of behavior and performance. The report identified five sets of issues for which research was required: environmental, psychophysiological, individual, interpersonal, and organizational. Such research should be interdisciplinary and conducted in ground-based analogue settings as well as in flight. Second, the efficacy of existing countermeasures (screening and selection, training, monitoring, support) should be determined. Such countermeasure

NASA'S CURRENT RESEARCH PROGRAM IN BEHAVIOR AND PERFORMANCE

According to estimates provided by NASA, NRA funding for research on behavior and performance accounted for a total of $4.8 million in FY 1999, including an FY 1999 augmentation of $1.3 million for new projects (Table 9.1). This ranks among the various disciplines as the second highest in funding and in number of projects. Within the National Space Biomedical Research Institute (NSBRI), additional funding (much of it provided by NASA) for behavior and performance research amounted to approximately $1 million in FY 1999. There are 19 projects with FY 1999 funding that involve research on behavior and performance; 14 were funded under the auspices of a NASA Research Announcement (NRA), and 5 were funded through NSBRI. Of the 14 FY 1999 projects funded in response to an NRA, 10 projects were under the direction of other extramural investigators and 4 were under the direction of NSBRI investigators.

Many of the behavior and performance projects with FY 1999 funding involve studies that address issues identified in the *Strategy* report. Issues related to sleep and circadian rhythms were addressed by the greatest number (11 projects), followed by issues related to neurovestibular functioning (6 projects). Collectively, sleep and circadian rhythms and neurovestibular function were the primary emphases of 17 of the 19 behavior and performance projects with FY 1999 funding. In FY 1999, perception and

TABLE 9.1 Summary of FY 1999 Funding for Behavior and Performance Subdisciplines

Subdiscipline	NRA Total ($)	No. of Projects	NSBRI Total ($)	No. of Projects
Environmental issues	0	0	0	0
Psychophysiological issues	2,258,194	9	1,013,760	5
Sleep, circadian	1,498,805	6	1,013,760	5
Human physical performance[a]	759,389	3	0	0
Stress and emotion	0	0	0	0
Individual	1,026,000	4	0	0
Psychological issues[b]	1,026,000	4	0	0
Psychiatric issues	0	0	0	0
Interpersonal issues	223,000	1	0	0
Organizational issues	0	0	0	0
Subtotal	3,507,194	14	1,013,760	5
New tasks	1,276,840	—	—	—
Total	4,784,034	—	—	—

[a]This area was not specified as a high priority in the *Strategy* report.
[b]All of these projects pertain to perception and cognition.

cognition and group dynamics were the focus of one project each. Most of these projects are ground based. None of the FY 1999 projects appear to involve the use of animals.

In addition to funded research projects, several operational studies in the area of behavior and performance have been planned by medical operations personnel at Johnson Space Center (JSC), and one has been initiated. These include a select-in validation study for long-duration flight; a critical task analysis for on-orbit performance assessment; an individualized workload limits monitoring system; a fatigue monitoring system; a Phase III chamber protocol to evaluate operational methodologies and hardware for sleep monitoring in simulated ISS activities; data collection on sleep medication efficacy, sleep quality, and pharmacodynamic assessment of "PRN" (i.e., as needed) medications during spaceflight; and the development of a Spaceflight Fatigue Assessment Test, Spaceflight Cognitive Assessment Test, and Spaceflight Behavioral Assessment Test. Apart from the cognitive assessment test, which has been certified for ISS, all of these projects are unfunded and remain in the planning stages.

Some behavioral research has also been conducted at NASA Ames Research Center. Two of the 38 projects listed in the FY 1998 Life Sciences Task Book were devoted to ARC-based studies of neurovestibular adaptation during spaceflight (NASA, 1999). None of the projects funded for FY 1999 appear to be based at Ames Research Center (ARC).

With respect to the five discipline areas identified in the *Strategy* report, current research efforts are most responsive to the psychophysiological issues listed as priority areas. One-third of the projects listed under behavior and performance in FY 1999 respond to the need for ground-based and in-flight studies of sleep architecture during long-duration missions, including predictors of change in sleep quality and quantity; whether sleep deprivation is cumulative; how much sleep debt is necessary to produce an overall impairment of cognitive performance, mood, and interpersonal behavior; and whether reductions in sleep debt are associated with improved performance. These projects account for 60 percent of all funding (NRA and NSBRI) allocated for behavior and performance in FY 1999. Several currently-funded projects have focused on the development and implementation of psychophysiological instrumentation for the assessment of behavior and performance in flight. Most of the currently funded studies of neurovestibular function examine change and stability in individual physiological patterns in response to microgravity, and a few examine the applicability of measures of neurovestibular function to measures of behavior and performance in flight. However, it is unclear whether any of these studies have focused on change and stability in *individual* neurovestibular patterns as described in the *Strategy* report. Furthermore, none of these studies have focused on physiological responses to psychosocial stressors.

In contrast to psychophysiological issues, none of the currently funded projects explicitly address organizational issues identified in the *Strategy* report, including the influence of organizational culture and mission duration on behavior and performance and the requirements for effective management of long-duration missions as they relate to task scheduling and timing and to distribution of authority and decision making. Lessons learned under actual field conditions in analogue settings (e.g., polar expeditions, military operations) suggest that these issues are likely to be among the most operationally significant in long-duration spaceflight.

One of the 19 FY 1999-funded studies is concerned with affective and cognitive responses to microgravity-related changes in perceptual and physiological systems and the monitoring of cognitive performance and affect in flight. Several aspects of cognitive performance, including psychomotor performance, information processing, short-term memory, and decision-making processes, have already been examined extensively. Since few, if any, cognitive performance decrements have been identified during short-duration missions, future research should begin to focus on potential decrements associated with long-term exposure to microgravity and physical or social monotony. No studies are currently examining the behavioral responses to perceived risks associated with the space environment (e.g., radiation, contamination of the ambient atmosphere, buildup of debris, use of breathing apparatus and space suits) or psychosocial predictors of the use and perceived importance of "personal" territories and individual strategies for coping with physical and social monotony.

Within the discipline area of individual issues, the use of certain coping strategies during long-duration missions, the association between personality characteristics and performance criteria, the relationship between self-reports and external (i.e., performance-related and physiological) symptoms of stress, and individual and mission-related predictors of postflight changes in personality and behavior are not being addressed at present by NASA-funded research. Projects that relate to the monitoring of cognitive performance and physiological indicators of performance address the relations between self-reports and external symptoms of stress and the effect of psychosocial stressors on cognitive performance, but only in a marginal fashion.

Only one of the FY 1999-funded projects addresses interpersonal issues related to the influence of different crew composition (i.e., in terms of personality type, gender, culture, language, occupation, and career motivation) on crew tension, cohesion, and performance during the mission. This project is consistent with *Strategy* report recommendations and will also address ground-crew interactions with astronauts and cosmonauts aboard the ISS, examining the impact of crew tension and dysphoria on crew-ground communication; the impact of ground-crew communication on crew cohesion versus task performance; and conditions that affect the distribution of authority, decision making, and task assignments between space crews and members of ground control.

In addition to the extramural activities described above, intramural research on behavior and performance at both JSC and ARC is ongoing, but largely operational in scope. The development of a critical path for research and countermeasure development at JSC includes four areas specifically related to behavior and performance: psychological adaptation (operational psychology), human-to-system interface, sleep and circadian assessment, and behavioral medicine. Priority has been placed on determining the following: (1) What fundamental behavioral stressors will most likely affect crew performance, both individual and team? (2) What information management and communication systems will best support the crew's ability to exchange information, learn and maintain proficiency on critical tasks, and meet mission objectives? (3) What are the acute and long-term effects of exposure to the space environment on biological rhythmicity and on sleep architecture, quantity, and quality? (4) What models of behavioral health and task performance best predict problems and provide guidelines for effective treatment of illness during a mission? In general, these priority areas exhibit a high degree of correspondence with the behavior and performance priorities listed in the *Strategy* report and reflect current directions in the extramural research arena.

In summary, current NRA- and NSBRI-funded research activities in the field of behavior and performance are consistent with the *Strategy* report recommendations for research on neurobiological mechanisms underlying the effects of physical environmental stressors on cognitive and psychophysiological measures of behavior and performance. Most of this research is concerned with the characteristics of sleep and circadian rhythms and with changes in cognition and perception related to alteration of neurovestibular function on long-duration missions. Although some research is being conducted currently on the psychosocial mechanisms underlying the effects of isolation and confinement on cognitive, affective, and psychophysiological measures of behavior and performance, the level of effort in addressing environmental, individual, interpersonal, and organizational issues is not consistent with the priority placed on these issues in the *Strategy* report.

PROGRAMMATIC BALANCE

Balance of Subdiscipline Areas

Current research efforts reflect considerable emphasis on sleep and circadian rhythms, performance related to neurovestibular function, psychophysiological monitoring, and cognitive performance on short-duration missions. In contrast, relatively little attention has been devoted to behavioral health and psychosocial adaptation, issues that could significantly impact the success of future long-duration missions. Current research efforts place too little emphasis on the influence of environmental, individual, and interpersonal factors and no emphasis whatsoever on organizational factors identified in the *Strategy* report as likely to affect behavior and performance during long-duration spaceflight.

This imbalance is reflected notably in the efforts of investigators currently affiliated with the NSBRI. The Human Performance Factors, Sleep, and Chronobiology Task Group is concerned almost exclusively with psychophysiological aspects of performance, concentrating on studies of sleep and chronobiology. The NSBRI External Advisory Council identified a gap in current research efforts directed at behavioral issues and recommended that greater emphasis be devoted to research and countermeasure development in neurobehavioral and psychosocial health. To this end, the NSBRI sponsored a workshop in July 1999 devoted to the task of identifying research priorities in this area. The workshop report identified six interrelated themes that define the range of factors critical to optimizing human performance during long-duration spaceflight: (a) biological mechanisms of neurobehavioral dysfunction; (b) cognition and performance; (c) individual factors; (d) team and interpersonal factors; (e) organizational, cultural and management factors; and (f) pharmacology in space. All six of these themes are consistent with those identified in the *Strategy* report.

Nevertheless, greater efforts are required to encourage the integration of extramural and intramural behavior and performance research, as well to propose and conduct research in areas related to individual, interpersonal, and organizational issues. The need for these efforts is reflected in the list of funded projects responding to the 1998 NRA (98-HEDS-02) "Gravitational Biology and Ecology and Biomedical Research and Countermeasures Programs." Only three of the funded projects appear to relate to behavior and performance; two of these concern sleep and circadian rhythms, and one concerns sensorimotor function. NASA's responsiveness to this need is reflected in the identification of specific psychological and psychiatric issues as topics for investigation in the 1999 NRA (99-HEDS-03) "Biomedical Research and Countermeasures."

Balance of Ground and Flight Investigations

Approximately one-third of current NASA-funded research in behavior and performance involves experiments and data collection activities conducted in flight. This research is essential to understanding fundamental physiological and psychological mechanisms underlying behavior and performance during extended spaceflight. Ground-based investigations in laboratories and analogue settings (i.e., isolation chambers, polar expeditions) are also critical to understanding these fundamental mechanisms. Moreover, they offer a cost-effective alternative to conducting similar studies in flight, with fewer logistic demands and greater sample sizes. For this reason, the current distribution between ground and flight investigations appears to be a reasonable one. Although current countermeasure testing and validation in flight is limited, such investigations will become increasingly important as the operational demands for valid and reliable interventions increase with the involvement of larger numbers of astronauts on long-duration assignments in space.

The current distribution between ground-based and in-flight studies also appears to be quite reasonable from the perspective of current constraints on conducting biomedical research in general, as well as behavior and performance research with NASA astronaut personnel during scheduled flights, and is consistent with *Strategy* report recommendations. However, it remains uncertain whether the current balance between ground and flight investigations is likely to continue. The construction phase of the International Space Station (ISS) will leave little opportunity for conducting in-flight experiments related to behavior and performance. Once the ISS is completed, biomedical research priorities must compete with commercial development and fundamental studies of microgravity for funding and personnel time. There has been no firm commitment to provide any of the resources necessary to conduct investigations to address these issues in the Human Research Facility racks proposed for the ISS.

Emphasis Given to Fundamental Mechanisms

Basic research designed to elucidate fundamental physiological mechanisms underlying such aspects of behavior and performance as sleep and circadian rhythms and sensorimotor and neurovestibular functioning accounts for most of the currently funded research in this area. These studies give appropriate emphasis to understanding fundamental mechanisms and demonstrate the need for such an understanding as a prerequisite to countermeasure development, testing, and validation. The same cannot be said of current research activities related to psychosocial issues. Most of these studies are observational in nature and lack the experimental design necessary to elucidate fundamental mechanisms. The *Strategy* report identified several issues that required research on fundamental mechanisms underlying relevant issues of behavior and performance during long-duration spaceflight. These include risk perception under conditions of isolation and confinement, interpersonal dynamics of small groups, crew tension and cohesion, the role of state versus trait characteristics in adapting to isolated and confined environments, and the causal links between alterations in the HPA axis, cognition, and affect. Greater effort is required to move beyond the application of broad-based models of behavior to the identification and support of empirically based advances in psychosocial theory.

Utilization and Validation of Animal Models

Although the use of animal models to understand certain fundamental physiological and psychosocial mechanisms related to behavior and performance in space is necessary and important, no specific mention of the use of animal models is made in the *Strategy* report. Some of the ground-based research on sleep and circadian rhythms has relied on animal models, and a few projects have attempted to validate these models under actual spaceflight conditions. Use of ground-based animal models to elucidate physiological mechanisms underlying behavior appears to be well developed and requires no additional emphasis beyond current levels. To date, there has been no attempt to develop animal models to elucidate psychosocial mechanisms, either on the ground or under actual spaceflight conditions. Such models are considered unnecessary and of questionable value at the present time.

In summary, in comparison to the *Strategy* report, current NASA-funded research places a disproportionate emphasis on sleep and circadian rhythms and on neurovestibular-sensorimotor decrements in perception and cognition and too little emphasis on the influence of environmental, individual, and interpersonal factors. There is no emphasis on organizational factors, identified in the *Strategy* report as likely to affect behavior and performance on long-duration spaceflight. The current balance of ground-based and in-flight studies is appropriate and understandable given the hiatus of in-flight opportunities during the construction phase of the International Space

Station. However, the *Strategy* report recommended that in-flight biomedical studies be accorded much higher priority once the construction phase is completed. The current program of NASA-funded research is also not consistent with the *Strategy* report recommendations for ground-based and in-flight studies of the fundamental mechanisms underlying psychosocial issues.

DEVELOPMENT AND VALIDATION OF COUNTERMEASURES

The high priority accorded by the *Strategy* report to the testing and validation of existing behavior and performance countermeasures and to the development, testing, and validation of new countermeasures is reflected in several recently published NASA operations documents (NASA, 1994, 1996, 1997a, 1998a,b). These documents outline current and proposed requirements for astronaut screening and selection, preflight training, in-flight monitoring and support, and postflight observation and support. They also identify areas in need of further investigation. For example, methods of validation for ISS crew selection and assignment have yet to be resolved.

However, there is very little evidence among current projects funded by NASA Life Sciences of research in the development, testing, and validation of countermeasures described in the *Strategy* report that are unrelated to sleep and circadian rhythms or to neurovestibular function. None of the FY 19999 NRA-funded projects listed under behavior and performance concern themselves primarily with countermeasure testing and validation, although the majority of projects have implications for countermeasure development. Four of the five NSBRI-funded projects are concerned with prevention and treatment of sleep- and circadian rhythm-related performance decrements. None of the recommendations for countermeasure development, testing, and validation listed in the *Strategy* report in the area of environmental, individual, interpersonal, and organizational factors appear to be the focus of current NASA-funded projects. Clearly, greater effort is required in countermeasure development, testing, and evaluation.

Consistent with *Strategy* report recommendations, the Countermeasures Task Force recommended that high priority be given to a critical review of analogue studies to identify psychological characteristics of successful individuals and crews, the natural history of psychosocial adaptation to extreme environments, identification of the types of problems that occur during deployment, and outcomes of attempted intervention and prevention measures (NASA, 1997b). Only three of the behavior and performance studies listed in the FY 1998 Task Book are in accordance with this recommendation. The *Strategy* report and the Countermeasures Task Force report also recommended that priority be given to reviewing the Russian experience with screening and selection, training and support, in-flight monitoring and support, and postflight monitoring and interventions. None of the current NASA-funded research projects address this recommendation. Also recommended in the Countermeasures Task Force report was a follow-up of the Rose et al. (1994) study of predictors of astronaut performance; research that improves our ability to assemble optimally functioning crews; and research on the best methods and protocols for training individuals and crews in conflict management and resolution, communication skills, cross-cultural awareness, team maintenance, and stress management.

Consistent with *Strategy* report recommendations, several behavior and performance countermeasures have been developed and implemented by the Medical Sciences Division at JSC, despite not having been tested and validated. Several others have been proposed, but remain in the planning stages. Current psychological adaptation countermeasures include select-out and select-in procedures for astronauts, family support office, preflight training in long duration mission and cross-cultural issues, in-flight psychological monitoring and support, and postflight debriefings. Proposed countermeasures include select-in procedures for the Expedition Astronaut Corps, individual and team field training and assessment, individual training support and consultation, crew conference plans, and confined operations training. Current human systems interface countermeasures address issues such as astronaut food preferences, aesthetics of work areas and crew quarters, and work schedules at all three stages of a mission (preflight, in flight, postflight). Proposed countermeasures include the development of on-orbit tools for assessment and retraining of critical skills; preflight assessment of optimal learning, workload, and limit assessment protocols; critical task analyses; and ergonomic fit to ISS requirements. Current sleep and circadian countermeasures include sleep self-report monitoring, shifting schedules to accommodate optimal sleep schedules, and administration of sleep medications in flight. Proposed countermeasures include sleep deficit and circadian rhythm monitoring, preflight assessment of fatigue limits, preflight training for individual fatigue countermeasures, and on-orbit monitoring of fatigue scale. Current behavioral illness countermeasures include select-out procedures for the astronaut corps, behavioral medicine care for astronauts and families, crew medical officer preflight training, in-flight monitoring and care, the Spaceflight Cognitive Assessment Tool, the private medical conference with the flight surgeon, and the administration of medications. Proposed countermeasures include an annual behavioral examination, select-out procedures for the Expedition Astronaut Corps, development of individual performance plans, preflight assessments of mood and stress, behavioral medicine meetings with astronauts and families preflight and postflight, and the Spaceflight Behavioral Assessment Tool.

As with the biomedical and operational research efforts described above, the development, testing, and validation of these countermeasures by JSC personnel are not supported by research programs but are funded almost exclusively from operational resources. However, there has been a growing voice within the Astronaut Program Office for the development and implementation of many of these countermeasures under the auspices of an Expedition Astronaut Training Program. The objectives of the program would be to train astronaut personnel assigned to tours aboard the International Space Station and other long-duration missions in three areas: self-care and self-management, leadership, and teamwork. The program would include both classroom and field exercises. Such a program would respond to several of the recommendations for development, testing, and validation of individual and interpersonal countermeasures listed in the *Strategy* report.

To ensure greater effort in countermeasure development, testing, and validation, as well as greater balance within the behavior and performance discipline areas in general, researchers should have exposure to operational needs and constraints. This will facilitate the design of studies that are likely to be relevant to the goals and objectives of the NASA community, particularly those that relate to countermeasures. Similarly, operations personnel have to call on extramural investigators for their expertise in addressing issues of operational relevance. The integrated product teams (IPTs) are designed to foster this collaboration between research and operations. However, the Behavior and Performance IPT has no current funding. The IPT mechanism for developing collaborative relationships among investigators and for integrating diverse research activities related to behavior and performance should be supported and strengthened. In addition, mechanisms for the interaction of these IPTs with personnel at NASA headquarters should be developed. Such efforts would facilitate preparation of RFAs, development and modification of research priorities, and dissemination of program activities.

In summary, with the exception of issues related to sleep and circadian rhythms or to neurovestibular functioning, current efforts leading to the development, testing, and validation of countermeasures are funded largely from the operations budget at JSC or are at the preliminary stages in NRA- and NSBRI-funded projects. None of the recommendations for countermeasure development, testing, and validation listed in the *Strategy* report in the area of environmental, individual, interpersonal, and organizational factors appear to be the focus of current NASA-funded projects. The committee recommends that greater effort be made in these activities, consonant with the critical importance of these tasks to the success of long-duration missions.

EPIDEMIOLOGY AND MONITORING

The capacity to monitor and assess psychological status preflight, in flight, and postflight was identified in the *Strategy* report as critical to the foundation of an effective and comprehensive program of research in the fundamental mechanisms underlying behavior and performance, as well as the development, testing, and validation of countermeasures designed to enhance performance effectiveness and reduce performance decrements. Consistent with *Strategy* report recommendations, the capacity for monitoring psychological status in flight is listed as a requirement in the ISS Medical Operations Requirement Document (MORD) (NASA, 1998a). However, there is no current requirement for pre- or postflight monitoring, nor is there a plan to systematically analyze and interpret the in-flight data to be collected. These activities offer a relatively noninvasive means of collecting data that are critical to the tasks of elucidating fundamental neurobehavioral and psychosocial mechanisms and of testing and validating countermeasures. An ongoing program of epidemiology and monitoring would also help to integrate research and operational activities in a manner that would be cost-effective, nonredundant, and likely to stimulate intellectual synergy and institutional collaboration.

To date, there has been no concerted research effort designed to quantify decrements in behavior and performance during spaceflight as specified in the *Strategy* report. Similarly, there has been no consistent effort to monitor such behavior in flight. Consequently, an understanding of the distribution of performance decrements among flight crews, particularly on long-duration missions, and of the factors contributing to an increased risk of such decrements remains largely unknown. Thus, high priority should be given to the development and implementation of such a plan, as well as to the expansion of monitoring activities to the preflight and postflight phases of a mission.

SUPPORT OF ADVANCED TECHNOLOGIES

Several new advanced technologies and methodologies have been supported by the extramural research program in behavior and performance. These include the adaptation of quantitative techniques for the analyses of cultural consensus and social dynamics for use in studies of small crews in isolated and confined environments; the development of statistical models to reliably detect in near real time significant alterations in the circadian physiology of individual subjects; the use of new techniques for the content analysis of qualitative data; and the development of automated data collection protocols for submitting psychosocial data from remote locations to investigators via the Internet. Several currently funded projects have already made substantial advances in the development of new, noninvasive techniques for monitoring cognitive, affective, and psychophysiological parameters of behavior and performance in flight, as recommended in the *Strategy* report. This includes the Performance Assessment Workstation to monitor cognitive performance and the use of neurophysiological measures (electroencephalographs) to monitor changes in cognitive status associated with environmental stressors.

The use of isolation chambers such as the new "BIO-Plex" facility at JSC offers the potential to study behavior and performance under conditions of extended isolation and confinement. Designed for the purpose of evaluating large-scale bioregenerative planetary life support systems with human test crews for long durations, BIO-Plex represents a high-fidelity test facility for the development of advanced technologies and methodologies for monitoring individual and interpersonal behavior, as well as for studies of fundamental mechanisms and countermeasure testing and validation. Its proximity to psychological and psychiatric operations personnel at JSC also provides an ideal context for fostering improved communication and collaboration among extramural investigators, NSBRI investigators and JSC operations staff in countermeasure development.

Currently funded NASA research in the area of behavior and performance appears to be making significant progress in the development of advanced technologies necessary for conducting noninvasive research in flight.

SUMMARY

Behavioral issues are likely to assume greater importance as missions in space grow in frequency and duration. The need to improve our understanding of fundamental neurobehavioral and psychosocial mechanisms and to develop, test, and validate countermeasures that optimize performance and minimize the performance decrements likely to occur during long-duration spaceflight has been amply demonstrated by past experience. At the present time, NASA-funded research programs only partially address this need.

REFERENCES

National Aeronautics and Space Administration (NASA). 1994. Astronaut Selection Policy and Procedures Manual of the NASA Behavioral Health and Performance Program. JSC Document No. 26887. Houston, Tex.: NASA.

NASA. 1996. Spaceflight Health Requirements Document (SHRD). JSC Document No. 26882. Medical Operations Branch. Houston, Tex.: NASA.

NASA. 1997a. Postflight Rehabilitation Program. Space and Life Sciences Directorate. JSC Document No. 27050. Houston, Tex.: NASA.

NASA. 1997b. Task Force on Countermeasures: Final Report. Washington, D.C.: NASA.

NASA. 1998a. International Space Station Medical Operations Requirements Document (ISS MORD): Baseline. SSP Document No. 50260. International Space Station Program Medical Operations Space and Life Sciences Directorate. Houston, Tex.: NASA.

NASA. 1998b. Astronaut Medical Evaluation Requirements Document (AMERD). JSC Document No. 24834. Space and Life Sciences Directorate. Houston, Tex.: NASA.

NASA. 1998c. The International Space Station Research Plan: An Overview. Washington, D.C.: NASA.

NASA. 1999. Life Sciences Task Book, FY 1998. Washington, D.C.: NASA.

National Research Council (NRC), Space Studies Board. 1998. A Strategy for Research in Space Biology and Medicine in the New Century. Washington D.C.: National Academy Press.

Rose, R.M., L.F. Fogg, R.L. Helmreich, and T.J. McFadden. 1994. Psychological predictors of astronaut effectiveness. Aviat. Space Environ. Med. 65:910-915.

10

Setting Priorities in Research

The *Strategy* report (NRC, 1998) considered the question of overall priorities for NASA-supported research in the next decade, taking into account budgetary realities and the need for clearly focused programs. In the biomedical arena, the report recommended that highest priority be given to research aimed at understanding and ameliorating problems that may limit astronauts' ability to survive and/or function during prolonged spaceflight. Such studies should include basic and applied research and ground-based investigations as well as flight experiments. It is clear that such problems must be identified and adequately solved or mitigated before long-term human spaceflight can be considered feasible.

It was further recommended that NASA programs focus on aspects of research in which NASA has unique capabilities or those that are underemphasized by other agencies. Six issues in biomedical research were identified as the most important for ensuring astronaut health, safety, and performance during and after long-duration spaceflight. The committee recognized that until the research facilities of the ISS are completed and can be fully utilized, NASA-supported research will necessarily be directed largely to ground-based investigations designed to frame critical hypotheses that can later be tested in space. Thus, implementation of recommendations for research that requires long-duration flight experiments is necessarily delayed until such opportunities reopen, although interim availability of a dedicated Shuttle flight could provide significant advances in some of the designated areas.

The following sections summarize the degree to which current programs focus on the six areas of research considered of highest priority. Details are presented in the disciplinary chapters.

LOSS OF WEIGHT-BEARING BONE AND MUSCLE

In the *Strategy* report, priority was given to five recommendations, of which two were directed primarily to ground-based research:

1. Studies that provide mechanistic insights into the development of effective countermeasures for preventing bone and muscle deterioration during and after spaceflight; and

2. Use of ground-based model systems, such as hindlimb unloading in rodents, to investigate the mechanisms of changes that reproduce in-flight and postflight effects.

Mechanistic studies and exploitation of ground-based animal models are indeed being emphasized by NASA in both disciplines. The growing use of genetically modified mice (knockout and transgenic) is of particular interest and importance, although flight experiments will be required to validate mechanistic hypotheses so obtained. Caging for mice suitable for Shuttle and International Space Station (ISS) flight experiments is now available.

A third recommendation of the *Strategy* report, to investigate the relationship between exercise activity and protein-energy balance in flight, will ultimately require flight experiments. However, preliminary ground-based studies will be important antecedents to flight and have recently been started.

Two additional recommendations—to obtain a database on the course of spaceflight-related bone loss and its reversibility in humans, and to establish hormone profiles on humans before, during, and after spaceflight—cannot be addressed until appropriate flight opportunities return.

VESTIBULAR FUNCTION, THE VESTIBULO-OCULAR REFLEX, AND SENSORIMOTOR INTEGRATION

The three recommendations of the *Strategy* report for high-priority research in the areas of vestibular function, the vestibulo-ocular reflex, and sensorimotor integration all involved extensive studies in spaceflight. Preliminary investigations were carried out on Neurolab related to the recommendation for in-flight recordings of signal processing following otolith afferent stimulation. However, the highest-priority recommendation, to determine the basis for the compensatory mechanisms on Earth and in space and evaluate whether the mechanisms are the same, has not yet been addressed. Similarly, studies to determine the effects of microgravity on the adaptation of the vestibulo-oculomotor system to sensory perturbations will require spaceflight.

In a parallel section defining high priorities for research in fundamental gravitational biology, the *Strategy* report recommended functional magnetic resonance imaging (fMRI) studies on astronauts pre- and postflight to determine the effects of microgravity on neural space maps. Initiation of such studies must also await the availability of appropriate flight opportunities.

ORTHOSTATIC INTOLERANCE UPON RETURN TO EARTH GRAVITY

High-priority issues included determination of the mechanisms underlying inadequate total peripheral resistance during postflight orthostatic stress; extension of current knowledge of cardiovascular adjustments to long-duration exposure to microgravity; reevaluation and refinement of existing countermeasures; and development of methods for referencing intrathoracic vascular pressures to systemic pressures in microgravity. These studies also require flight opportunities. Although the ability to conduct experiments on the ISS will ultimately be required, much of the necessary knowledge could be obtained from short-term Shuttle-based flight and postflight studies.

RADIATION HAZARDS

Two of the four high-priority recommendations in this area require improved availability of high-energy and high-Z accelerator facilities:

1. Determine the carcinogenic risks following irradiation by protons and high-atomic-number, high-energy (HZE) particles; and
2. Determine if exposure to heavy ions at the level that would occur during deep-space missions of long duration poses a risk to the central nervous system.

These issues are being addressed by National Space Biomedical Research Institute (NSBRI) investigators, and new facilities at Loma Linda University for proton studies and at Brookhaven for heavy ions will provide greatly improved access to investigators for relevant studies. Both areas are given priority in the 2000 NASA Research Announcement (NRA) for biomedical research and countermeasures.

One recommended area of study, to determine whether combined effects of radiation and stress on the immune system in spaceflight might have additive or synergistic effects on host defenses, has not been implemented, although preliminary ground-based animal studies could give important insights into this question (Todd et al., 1999).

The remaining recommendation, to determine how selection and design of the space vehicle affect the radiation environment, goes beyond the boundaries of the biomedical research program and is not considered in the present report.

PHYSIOLOGICAL EFFECTS OF STRESS

The *Strategy* report contained a single high-priority recommendation to examine the role that the host response to stressors plays in alterations in host defenses by investigation of the interactions between the neuroendocrine hypothalamic-pituitary-adrenal (HPA) axis and the immune system during spaceflight. Flight opportunities are lacking for the near term, and although the area has attracted attention under the aegis of other agencies, ground-based studies have not previously received significant attention in NASA programs. However, the 2000 NRA called explicitly for proposals to evaluate effects of stressors on physiological function and emphasized studies using integrated approaches.

PSYCHOLOGICAL AND SOCIAL ISSUES

The *Strategy* report gave highest priority to a recommendation for interdisciplinary research on the neurobiological and psychosocial mechanisms underlying the effects of environmental stressors on cognitive, affective, and psychophysiological measures of behavior and performance. The majority of current NASA-supported research concerns neurobiological and neurobehavioral aspects of stress-related effects, with an emphasis on perturbing effects of spaceflight on circadian rhythms and sleep patterns. Until recently, less attention has been paid to psychosocial stressors and their effects; however, the 1999 and 2000 NRAs gave explicit emphasis to these issues. Very little of the research to date has had a significant interdisciplinary focus, although the NSBRI programs are beginning to move in this direction. The International Life Sciences Working Group has also recommended a greater emphasis on interdisciplinary approaches.

A second high-priority recommendation was for interdisciplinary research to evaluate the efficacy of existing countermeasures, develop more effective replacements for those that are deemed inadequate, study the use of psychophysiological countermeasure implementation, and determine the effects of the spaceflight environment on the kinetics and efficacy of psychopharmacological medications. With the exception of the circadian and neurovestibular systems, countermeasure evaluation and development received little attention in the past. However, the 2000 NRA, in soliciting research on the development of predictive tools for the assessment and support of psychological well-being, has begun to redress this imbalance.

REFERENCES

National Research Council (NRC), Space Studies Board. 1998. A Strategy for Research in Space Biology and Medicine in the New Century. Washington, D.C.: National Academy Press.

Todd, P., M.J. Pecaut, and M. Fleshner. 1999. Combined effects of space flight factors and radiation on humans. Mutat. Res. 430:211-219.

11

Programmatic and Policy Issues

The 1998 *Strategy* report raised concerns in the program and policy arena, including issues relating to strategic planning, the conduct of space-based research, and utilization of the International Space Station (ISS), as well as mechanisms for promoting integrated and interdisciplinary research and collection of and access to human flight data (NRC, 1998). The following sections summarize concerns remaining in these areas. In addition, the chapter describes several important issues having to do with countermeasure testing and validation and the role of the Space Medicine Program in human research that came to the committee's attention during the course of the present study.

INTERNATIONAL SPACE STATION: UTILIZATION AND FACILITIES

The adequacy of research facilities on ISS remains a serious concern. One new issue has arisen with respect to the variable force centrifuge. The committee understands that the amount of power available for the centrifuge may be substantially below initial specifications. It is not clear to the committee whether and to what extent the usefulness of this crucial instrument may be compromised by possible power limitations.

Previous concerns about the effects of cuts in the utilization budget on the capacity of ISS to support high-quality research remain in force. Timetables for incorporation of relevant research facilities and equipment have been eroded by continuing budget cuts and delays in assembly schedules for research facilities, and the confidence of the user scientific community continues to be eroded by the perceived potential for downgrading of research capabilities and budgets. The most recent assembly sequence available to the committee (June 1999) is summarized in Table 11.1. Limited capability for human research is scheduled to begin in mid-2000, with further addition of instrument racks and a −80 °C life sciences freezer by mid-2001. The second human research facility is to be added in March 2001. However, habitat holding facilities and the life sciences glove box, necessary for animal and tissue culture experiments, will not be available until 2003, and the variable speed centrifuge is not scheduled until mid-2004.

TABLE 11.1 International Space Station Assembly Sequence for Biomedical Research Facilities

Date	Flight	Element Deployed
June 2000	5a.1	Human Research Facility-1
January 2001	UF-1	Minus Eighty Degree Freezer
March 2002	12A.1	Human Research Facility-2
February 2003	UF-3	Habitat Holding Rack-1
		Life Sciences Glovebox
September 2003	UF-5	Habitat Holding Rack-2
August 2004	UF-7	Centrifuge Accommodation Module

NOTE: Based on Revision E assembly sequence.

Questions also remain about the role of Russian cosmonauts in the conduct of biomedical research, especially during the early phases of ISS utilization. Issues that must be clarified include the nature and extent of the training these crew members will receive for the conduct of ISS-based research protocols, their commitment to participate as subjects in human studies, and the nature and adequacy of postflight longitudinal follow-up to such human research studies.

In conclusion, the adequacy of the life sciences research facilities that will actually be in place on the ISS at its final build-out remains an issue of serious concern. Possible design changes, the mounting delays in utilization timetables, and the perceived potential for downgrading of research facilities and budgets have continued to erode the confidence of the user scientific community. Important questions also remain about the role of Russian cosmonauts in the conduct of biomedical research, especially in the early phases of ISS utilization.

COUNTERMEASURE TESTING AND VALIDATION

The availability of effective countermeasures against the deleterious effects of spaceflight on astronaut health and performance will be an increasingly critical issue as longer-duration flights become the norm on the ISS and beyond.

Development of effective, mechanism-based countermeasures requires three well-integrated phases: (a) basic research, ground-based and in flight, to identify and characterize mechanisms of spaceflight effects; (b) testing and evaluation of proposed countermeasures, to determine their efficacy in ground-based models of the flight environment; and (c) validation of promising countermeasures by well-designed clinical studies in flight as well as pre- and postflight. Maintenance of a longitudinal database documenting relevant physiological and performance parameters in the presence and absence of the given countermeasure will be crucial to validating efficacy.

Basic research whose ultimate goal is the development of improved countermeasures is a primary mission of the National Space Biomedical Research Institute (NSBRI) and a major component of the life sciences NASA Research Announcement (NRA) program overall, while Johnson Space Center (JSC) has primary responsibility for ground-based testing and evaluation of proposed countermeasures. There has been no well-established, standard procedure whereby newly proposed countermeasures can gain access to the evaluation pipeline, and well-defined, published criteria for accepting candidate countermeasures for testing appear to be lacking. A defined process for carrying out such testing and evaluation is currently under development at JSC but has not yet been implemented. It is essential that the process be readily accessible to all investigators, extramural as well as intramural, and that criteria for acceptance into the testing program be clearly defined.

Just as a drug must be proven safe and effective before being used clinically, a countermeasure must be validated to establish that it prevents or ameliorates the problem it addresses and does not show significant unanticipated side effects in physiological systems. Initially, the Detailed Supplemental Objective (DSO) program, based at JSC, provided the mechanism for studies to validate countermeasures in flight. Its successor, the Extended Duration Orbiter Medical Project (EDOMP), was established when Shuttle flights became 16 days or longer. EDOMP has continued the practice of including testing as a secondary mission objective, with the goal of optimizing crew performance. In general, however, the practice of carrying out countermeasure validation as a secondary mission objective has not succeeded in providing definitive answers, even when tests are well designed, because of in-flight modifications or suboptimal conduct of protocols.

Countermeasure validation can take many forms and requires a firm definition when spaceflight countermeasures are discussed. The committee considered three types of validation for spaceflight countermeasures: (a) controlled trials in space, (b) comparison with historical controls, and (c) empirical observation of effects.

Controlled Trials in Space

The double-blind, placebo-controlled trial is the standard for many therapies in clinical practice and should be the standard whenever possible for spaceflight investigations. Several factors, however, have limited the number of these trials in space. First, the number of subjects is necessarily limited. The number of subjects required to show a statistically significant effect is based on a power calculation. For example, consider a study of bone loss in which two treatments are compared with a placebo and the end point is to detect a 1 percent difference in bone density. Based on published data on machine precision and measurement variability, it could take approximately 64 subjects per group to provide an 80 percent probability of seeing a significant effect at a 95 percent confidence level. Clearly, this would take years to complete on the Space Station, even if other confounding factors could be removed. The second limitation involves the environment and the variability of mission length, medication, exercise protocols, diet, and so forth, which cannot be controlled reliably over a series of missions with different goals. The third limitation is blinding. Not all treatments (exercise protocols, for example) can be evaluated in a double-blind fashion. Finally, the use of placebos is problematic. Depending on the nature of the clinical problem, operational pressures are likely to lead to preferring some treatment, even if unvalidated, rather than letting the problem develop uncountered.

To deal with these limitations, each countermeasure and problem has to be scrutinized carefully. Can a meaningful trial be performed with the sample size available, or is only a less rigorous validation possible and sufficient? Will an unblinded trial or a comparative trial without placebo be acceptable? Recent advances in statistical theory have made the conduct of small-sample clinical trials more feasible. Use of expert statistical consultants on a continuing basis to assist in the design and analysis of countermeasure studies would be an important addition to the overall countermeasure program. In addition, ground-based trials in analogue settings provide important information on the potential effectiveness of candidate countermeasures in spaceflight. Only the most critical countermeasures should have the highest level of validation, but if this is the required approach, all of the resources needed to do the study correctly must be provided.

Use of Historical Data

Although a placebo-controlled trial may not be possible, a comparison of the effects of the countermeasure on the variable of interest with historical data may still provide some level of validation. An example might be the effects of a drug or treatment on urinary calcium excretion. This approach would provide some evidence that a drug or intervention is having a positive effect. Unfortunately, existing data are fragmentary and are not generally available to investigators. A rigorous, active program is required to collect and tabulate data from Shuttle and Space Station missions. Without an accurate and accessible database on the human responses to weightlessness and the incidence of various conditions, the historical approach cannot work. This means that the data collected on any mission are critical and may be essential for evaluating countermeasures in the future. A well-defined set of baseline data collected on all missions and available to investigators is crucial for the development and validation of optimally effective countermeasures.

Empirical Observation

Countermeasures have generally been implemented without in-flight validation. Some have been implemented with little or no evaluation, and others have been implemented based on effectiveness in ground-based studies or in analogue settings. It is likely that the pressures of operations and the difficulty of conducting controlled trials may continue to make this approach a frequent basis for implementation. However, for this level of validation to work, the ground-based analogue setting must have high fidelity to flight conditions, and a rigorous and consistent method of review is essential. The efficacy of the countermeasure must be monitored by continued, ongoing evaluation and assessment. Data should be collected to ensure that the problem is in fact being addressed and that other unanticipated effects are not occurring. For example, the fluid loading countermeasure against orthostatic intolerance has been implemented for spaceflight. Is this being well tolerated? Are there problems with gastric upset or vomiting due to the high osmotic load? What is the level of participation? Has the incidence of pre-syncope decreased? Without ongoing review, the countermeasure cannot be improved, problems will not be uncovered, and astronauts will be exposed to increased risk.

The committee recognizes the difficulties involved in proving that spaceflight countermeasures are effective. Validation, however, is essential and should be explicitly addressed for every proposed and implemented countermeasure. The main questions should be the following:

- Is this a countermeasure that requires validation in flight, and if so, how many subjects will be needed to prove the effect?
- Are appropriate data currently being collected that are likely to be useful for countermeasure development in the future?
- Is a rigorous countermeasure evaluation and assessment program in place to monitor any adopted countermeasure?

The need for effective countermeasures against deleterious effects of spaceflight on astronaut health and performance will become increasingly critical as longer-duration flights become the norm on the ISS and beyond. Development of effective, mechanism-based countermeasures requires three well-integrated phases: (a) basic research to identify and understand mechanisms of spaceflight effects; (b) testing and evaluation of proposed countermeasures to determine their efficacy; and (c) validation of promising countermeasures by well-designed clinical studies. Re-

cently NASA has begun to develop a standard procedure for the testing and evaluation of countermeasures, but this has not yet been implemented. It is essential that the process, once in place, be readily accessible to all investigators, extramural as well as intramural, and that criteria for acceptance into the testing program be clearly defined.

OPERATIONAL AND RESEARCH USE OF BIOMEDICAL DATA

As discussed above for countermeasure validation, historical data are essential to establish in-flight physiologic norms and the incidence of adverse events. For example, the mean urinary calcium excretion in space is high. If a drug were used to decrease calcium excretion, one way to measure its effectiveness would be to compare calcium excretion from people on the drug to an accurate historical mean. For long-duration flights, a series of medical evaluation requirements, the Integrated Testing Regimen (ITR) has been proposed. Many of these tests, such as blood analyses, body mass, and extravehicular activity (EVA) heart rates, will be important not just for the health and safety of the person providing the data, but also for clinical research intended to establish norms and trends to benefit future flyers. Clinical investigators should be involved in identifying the parameters to be tested.

Access is limited to these in-flight data, as well as to data collected in postflight longitudinal monitoring of astronaut health. Incomplete availability of human data to qualified investigators was highlighted as a major concern in the *Strategy* report. Individual data collected as part of the medical operations program are confidential. However, aggregate data, which cannot be attributed to an individual, are not. These data are critical for detecting trends and for developing and evaluating countermeasures. The rationale for selection of measurements to be made in flight and postflight is not clear, and should be held to the same high standard as the selection of research protocols. In fact, they should be treated as research data using the same rigor and control found in research studies. Data should be provided to the scientific community and reviewed periodically to ensure that a useful, accurate database is being established. The *Strategy* report recommended that expert scientific panels be employed to determine which data must be collected to serve research purposes appropriately. Long-duration human spaceflight is an ongoing research and development effort. The right to individual privacy and confidentiality is an established national policy; therefore, nonattributable medical data are essential to the future of crewed spaceflight and require careful oversight, review, and publication.

The Role of Medical Operations in Human Research and Countermeasure Validation

Role of the Crew Surgeon

The ISS Medical Operations Requirements Document proposes routine nominal health and fitness evaluations for crews during flight (NASA, 1998b). The crew surgeon will play a major role in defining crew member activities before, during, and after flight. The surgeon will participate in and provide oversight of the medical care of crew members during the launch, in-flight, and landing phases of the mission; perform medical certification; monitor medically hazardous training events; require biomedical baseline data collection; develop and implement in-flight countermeasures; review mission payload activities; establish in-flight time line and scheduling constraints; develop mission-specific aeromedical flight rules; and staff the Mission Control Center during flight. A mandatory period of postflight rehabilitation with milestones for return to 1 g normal health will be imposed. The high priority and demanding nature of this medical operations program leave little room for scientific investigation on humans. Given the command position of the crew surgeon, who is responsible for implementing and verifying countermeasures, rigorous training in clinical and basic research is recommended. Such training would greatly facilitate the communication and working relationship of investigators with NASA operational medicine programs during the flight investigation and would enhance scientific returns. In addition, better coordination in flight is necessary to minimize the deleterious effects of operational exigencies on the successful conduct and completion of research experiments.

Collection of Longitudinal Human Data for Research Purposes

The Astronaut Medical Evaluation Requirements Document requires extensive blood work on landing day, but minimal blood work later (NASA, 1998a). From a research perspective, the list of measurements and the sampling schedule have not been optimized to give the most useful information on in-flight development of problems and postflight recovery of normal physiological function. As recommended by the *Strategy* report, the definition of what and when measurements should be taken in flight and in longitudinal studies postflight (e.g., for the integrated testing regimens) requires input from experts in the relevant disciplines to optimize the return of health and science information. As long as the use of this information remains restricted and scientific input into what should be measured is lacking, the utility of this information in aiding the development of countermeasures will be severely curtailed. In addition, it will be important for NASA to foster the development of advanced technologies to provide automated noninvasive and minimally invasive in-flight and longitudinal monitoring for acquisition of the desired data.

Effects of Spaceflight on Drug Efficacy and Pharmacokinetics

There have been a number of largely anecdotal reports of altered drug efficacy or changes in pharmacokinetics in spaceflight. Although clinical pharmacology was not emphasized in the *Strategy* report, carefully designed clinical research is needed to determine whether there are significant changes in the effectiveness of drugs of interest, to define likely causes, and to develop effective countermeasures.

Availability of Stored Clinical Samples

A large bank of frozen clinical samples (urine, blood, etc.) has been saved over time and stored at JSC. If the quality of the samples has been maintained, these materials could be of significant interest and value for ongoing and future research purposes. However, the quality of some or all of the older material may have been compromised by freezer malfunctions over the years, and the issue of patient confidentiality as related to ongoing availability of samples to the investigator community has not been addressed. Efforts should be made to determine which sets of samples are of appropriate quality and to explore means of making them known and generally accessible to the relevant investigator community.

Data Archive

Timely completion of an archival research database was assigned high priority in the *Strategy* report. The Life Sciences Data Archive within the Program Integration Office at JSC is responsible for this activity, which is ongoing. However, the committee is concerned that uncertainties in funding for data entry have impeded the ability of the office to bring the archive up to date. The data archive is an important tool in making the results of NASA's biomedical investigations accessible to the research community, and NASA should consider completion of the project a priority.

In summary, access to in-flight biomedical data, as well as to longitudinal data collected in postflight longitudinal monitoring of astronaut health, is limited. The partial and incomplete availability of human data to qualified investigators was highlighted as a major concern in the *Strategy* **report and continues to be an issue. The committee urges that NASA explore ways in which these data and samples, collected in the past and future, can be made available to investigators. Additionally, steps are needed to ensure that future data collection includes measurements and sampling that have been optimized to give the most useful information on in-flight development of problems and postflight recovery of normal physiological function. The role played by the crew surgeon is especially critical to collection of these data, and rigorous training in both clinical and basic research is recommended as a requirement for the position.**

SCIENCE POLICY ISSUES

Support of Operational Research

Operational or applied research can be defined as research that is targeted to solve a specific, often narrowly defined, problem—for example, the optimal prebreathing protocol for EVAs. Operational research of this kind has generally been carried out intramurally, via the DSO or more recent EDOMP program. There is concern that in the context of peer review carried out under the life sciences NRA program, such research may sometimes be considered of lesser interest and priority than "basic" studies, even though the problem in question may have significant import for astronaut health, safety, and performance. However, extramural investigators will often bring important expertise and insight to issues requiring operational research and should be encouraged to carry out such studies. Focused NRAs designed to elicit proposals dealing with operational research issues, in conjunction with focused peer review groups, would provide a mechanism for the entire research community, intramural and extramural, to address areas of specific need more effectively. In addition, NASA should make a greater effort to acquaint applicants and investigators with the flight milieu as it relates to practical issues of space-based research protocols and expectations.

International Cooperation

The Division of Life Sciences is to be congratulated on its successful establishment of a process of international peer review with its European partners that provides scientific review of all investigator-initiated proposals. This is an important advance and a noteworthy example of international cooperation. The committee looks forward to the possibility that additional members of the international space community will enter the international peer review process in the future.

The recent establishment of an International Space Life Sciences Working Group is also a positive step toward ensuring coordination of research activities among the international partners in the ISS. In addition, however, the era of ISS construction and utilization, with its increased emphasis on international crews and operations, raises important issues with respect to acquisition and management of human data. Mechanisms are needed to ensure that protocols and facilities for pre- and postflight monitoring and testing are consistent across national boundaries. There have to be common criteria for evaluation and utilization of countermeasures and international cooperation in their development.

Integration of Research Activities

Roles of Program Constituents

With the advent of NSBRI, continued attention will be necessary to define the roles of the institute, NASA centers, NASA Specialized Centers of Research and Training (NSCORT), and the broader extramural biomedical community in the conduct of NASA's overall biomedical research program. The primary mission of NSBRI is to conduct basic research aimed at development of countermeasures, and the 1998 NSBRI Annual Report gives solid evidence of progress in establishing institute programs and initiating high-quality disciplinary and interdisciplinary research in many of the relevant areas of study. Interactions between NSBRI and JSC are still evolving. Effective collaborations appear to be in place between some intramural scientists and NSBRI members, but it is not entirely clear to what extent the goal of interaction is being met in practice. Indeed, the problem of coordination among and within the various program constituencies is a general one and requires continuing NASA oversight of the entire program and implementation of mechanisms, such as the recently initiated Biennial Biomedical Investigators' Workshop, to facilitate coordination of efforts.

A second significant concern noted in the *Strategy* report is the role of NSBRI in the overall biomedical NRA program. As the boundaries of NSBRI activities continue to expand, there is a growing need to delineate carefully the roles and responsibilities of the institute in relation to the broader NRA program. It is important for NASA to maintain a healthy NRA program as the primary mode for support of space-related biomedical research. The NRA program is the channel for support of space-related research for the entire community of investigators—extramural, intramural, and NSBRI affiliated. Exploration of novel ideas and approaches is best accomplished by maintaining access of the entire investigator community to NASA research programs. The research mission of NSRBI is focused on a limited number of areas, whereas the NRA program is more broadly based. In addition, since the NSBRI mission under current policy is limited to ground-based studies, a vigorous NRA program is also crucial to proper development of flight experiments. Thus, the potential impact of NSBRI funding on the overall biomedical research budget in the coming years should be carefully monitored and evaluated on a regular basis.

Intramural scientists are and will remain an essential component of the overall research program, both to provide expertise to the extramural principal investigator (PI) community in the utilization of unique research facilities and in the conduct of applied and operational research, and to provide an interface between the extramural PIs and NASA engineers in the development of experimental protocols and flight hardware. Intramural investigators operate under a number of constraints in their efforts to collaborate with extramural scientists. Investigators are given little institututional support or incentive to conduct research or to publish results in peer-reviewed journals. The heavy operational workload, attributed to staffing cuts, undermines the ability to do active research. The committee notes that very limited travel funds for intramural scientists have had a negative effect on their ability to attend and participate in meetings with their extramural peers. Such interactions are important both to the maintenance of the scientific knowledge base on which new research is based and to the reputation of NASA scientists as respected members of the research community. Consequently, intramural investigators risk becoming second-class members of research teams that include extramural principal investigators.

Integration of research activities and facilitation of collaborative and interdisciplinary research are dependent on open and effective communication among researchers over the entire program. Thus, the highly successful First Biennial Biomedical Investigators' Workshop, held in January 1999, which brought together all funded biomedical investigators, marked the beginning of an important new program (see NASA and USRA, 1999). The committee strongly supports its continuation and commends publication of the workshop proceedings as an especially valuable tool for the community.

Interagency Collaboration

Convincing testimony to the extent and value of collaborations between the Division of Life Sciences and the National Institutes of Health is provided by the success of Neurolab, as well as by jointly funded grants under the NRA and NSBRI programs. The National Science Foundation's Antarctic program provides a laboratory for behavioral research in isolated environments that is increasingly important as the era of prolonged occupation of the ISS approaches, and collaborations with the Department of Defense, the Department of Energy, and the Office of Naval Research have significant benefits for advanced technology development. The committee strongly supports such interagency collaboration, both as a means of stretching tight budgets for research and development and as a means of adding new scientific expertise to important problems of mutual interest.

In summary, operational research—that is, targeted research directed at solving a specific, well-defined problem—has generally been carried out by NASA intramural investigators. The larger extramural investigator community should also be encouraged to engage in this type of research, by the use of NRAs and peer review groups focused on issues in operational research.

The era of ISS construction and utilization, with increased emphasis on international crews and operations, raises important issues with respect to acquisition and management of human data. Mechanisms are needed to ensure that protocols and facilities for pre- and postflight monitoring and testing are consistent across national boundaries. There have to be common criteria for the evaluation and utilization of countermeasures and international cooperation in their development.

NASA funding for biomedical research is increasingly distributed among a diverse set of organizations and programs. These include the program of NASA Research Announcements, intramural investigators in NASA center science programs, the NSBRI, and NSCORT. NASA science benefits from the unique strengths of each of these program constituents. However, careful planning is required to delineate the roles, responsibilities, and appropriate funding levels for each; to ensure effective collaborations; and to integrate research findings. In particular, NASA should maintain a healthy NRA program as the primary mode for support of space-related biomedical research because of the advantages it offers in accessing the widest investigator community and exploring novel ideas and approaches.

REFERENCES

National Aeronautics and Space Administration (NASA). 1998a. Astronaut Medical Evaluation Requirements Document (AMERD), JSC 24834, Rev. A. Houston, Tex.: NASA.

NASA. 1998b. International Space Station Medical Operations Requirements Document (ISS MORD), Baseline SSP 50260. Houston, Tex.: NASA.

NASA and Universities Space Research Association (USRA). 1999. Proceedings of the First Biennial Biomedical Investigators' Workshop, January 11-13, 1999, League City, Texas. Houston, Tex.: NASA and USRA.

National Research Council (NRC), Space Studies Board. 1998. A Strategy for Research in Space Biology and Medicine in the New Century. Washington, D.C.: National Academy Press.

Appendixes

Appendix A

A Strategy for Research in Space Biology and Medicine in the New Century

Executive Summary

INTRODUCTION

The core of the National Aeronautics and Space Administration's (NASA's) life sciences research lies in understanding the effects of the space environment on human physiology and on gravitational biology in plants and animals. The strategy for achieving that goal as originally enunciated in the 1987 Goldberg report, *A Strategy for Space Biology and Medical Science for the 1980s and 1990s*,[1] remains generally valid today. However, during the past decade there has been an explosion of new scientific understanding catalyzed by advances in molecular and cell biology and genetics, a substantially increased amount of information from flight experiments, and the approach of new opportunities for long-term space-based research on the International Space Station. A reevaluation of opportunities and priorities for NASA-supported research in the biological and biomedical sciences is therefore desirable.

The strategy outlined in the Goldberg report had two main purposes: "(1) to identify and describe those areas of fundamental scientific investigation in space biology and medicine that are both exciting and important to pursue and (2) to develop the foundation of knowledge and understanding that will make long-term manned space habitation and/or exploration feasible."[2] To achieve these purposes, the Goldberg report identified four major goals of space life sciences:

"1. To describe and understand human adaptation to the space environment and readaptation upon return to earth.

"2. To use the knowledge so obtained to devise procedures that will improve the health, safety, comfort and performance of the astronauts.

"3. To understand the role that gravity plays in biological processes in both plants and animals.

"4. To determine if any biological phenomenon is better studied in space than on earth."[3]

These goals remain valid and form the basis of the present report.

NOTE: Reproduced from National Research Council, Space Studies Board, 1998, *A Strategy for Research in Space Biology and Medicine in the New Century*, National Academy Press, Washington, D.C., pp. 1-18.

Both the Goldberg report and the 1991 follow-up assessment, *Assessment of Programs in Space Biology and Medicine 1991*,[4] emphasized basic research and the importance of vigorous ground-based programs aimed at addressing the fundamental mechanisms that underlie observed effects of the space environment on human physiology and other biological processes. The present report strongly reemphasizes that strategy, and calls for an integrated, multidisciplinary approach that encompasses all levels of biological organization—the molecule, the cell, the organ system, and the whole organism—and employs the full range of modern experimental approaches from molecular and cellular biology to organismic physiology.

The sections that follow summarize the Committee on Space Biology and Medicine's priorities for NASA-supported research, its recommendations for high-priority research in individual disciplines, and its recommendations for overall priorities for NASA-sponsored research across disciplinary boundaries. The final section outlines significant concerns in the program and policy arena and offers related recommendations.

PRIORITIES FOR RESEARCH

Taking into account budgetary realities and the need for clearly focused programs, the highest priority for NASA-supported research in space biology and medicine in the new century should be given to research meeting of one of the following criteria:

1. *Research aimed at understanding and ameliorating problems that may limit astronauts' ability to survive and/or function during prolonged spaceflight.* Such studies include basic as well as applied research and ground-based investigations as well as flight experiments. NASA programs should focus on aspects of research in which NASA has unique capabilities or which are underemphasized by other agencies.

2. *Fundamental biological processes in which gravity is known to play a direct role.* As above, programmatic focus should emphasize NASA's capabilities and take into account funding patterns of other agencies.

A lower priority should be assigned to areas of basic and applied research that are relevant to fields of high priority to NASA but are extensively funded by other agencies, and in which NASA has no obvious unique capability or special niche.

HIGH-PRIORITY DISCIPLINE-SPECIFIC RESEARCH

Because the recommendations for research, and research priorities, in the discipline-specific chapters cover a wide range of fields relevant to space biology and medicine, the committee chose not to reproduce all of those recommendations in full in this executive summary. Instead the committee sought to capture the essence of what is recommended in Chapters 2 through 12, an approach that was best served by condensation, full quotation, or addition of supplemental detail as seemed useful to preserve the intent of the recommendations in their full form and context. The recommendations are numbered only in instances in which the committee considered that there was a clear priority order.

Cell Biology

Rapid advancement in the field of cell biology offers novel opportunities for studying the effects of spaceflight, including weightlessness, on cells and tissues. This possibility for progress stems both from developments in technology and advances in basic concepts of cell structure and function at the molecular level. Reasonable goals for the next period of NASA investigation are to clearly delineate the specific cellular phenomena that are affected by conditions of microgravity, to develop an understanding of the molecular mechanisms by which these changes are induced, and to begin to suggest strategies for countermeasures where indicated. Experience from previous in-flight and ground-based studies has highlighted certain pitfalls that must be avoided in the design and analysis of future experiments. Cellular systems should be emphasized that are known to be affected by gravitational force (e.g., bone, muscle, and vestibular systems in animals; gravitropic systems in plants) or by other aspects of the space environment (e.g., stress-induced phenomena). Consideration should be given to using molecular techniques for the analysis of gene expression and cell architecture and function, and to extending cell culture studies to the analysis of cellular physiology in intact tissues and whole organisms.

The committee makes the following specific recommendations for research in cell biology:

- General mechanisms of mechanoreception and pathways of signal transduction from mechanical stresses are areas of special opportunity and relevance for NASA life sciences. Studies of mechanisms of cellular mechanoreception should include identification of the cellular receptor, investigation of possible changes in membrane and cytoskeletal architecture, and analysis of pathways of response, including signal transduction and resolution in time and space of possible ion transients.
- Studies of cellular responses to environmental stresses encountered in spaceflight (e.g., anoxia, temperature, shock, vibration) should include investigation of the nature of cellular receptors, signal transduction pathways, changes in gene expression, and identification and structure and function analysis of stress proteins that mediate the response.
- The successful conduct of sophisticated cell biological experiments in space will require the development of highly automated and miniaturized instrumentation and advanced methodologies. NASA should work with the scientific community and industry to foster development of advanced instrumentation and methodologies for space-based studies at the cellular level.

Developmental Biology

The specific physiological systems in humans and animals for which gravity is likely to play a critical role in development and/or maintenance include the vestibular system, the multiple sensory systems that interact with the vestibular system, and the topographic space maps that exist throughout the brain. Major changes in perspective in recent years in the general field of developmental biology could greatly affect our ability to study and understand these systems. In particular, the use of saturation mutagenesis to identify genetic components of development, the recognition that molecular mechanisms are conserved across phylogeny, and the information provided by genome sequencing projects, have transformed basic developmental studies since the publication in 1987 of the Goldberg report.[5] In the present report the committee stresses the importance of two types of studies, those looking at life cycles and those examining development of gravity-sensing systems such as the vestibular system.

Complete Life Cycles in Microgravity

- The committee recommends that key model organisms be grown through two complete life cycles in space to determine whether there are any critical events during development that are affected by space conditions. Because no critical effects have been seen in model invertebrates, the highest priority should be given to testing vertebrate models such as fish, birds, and small mammals such as mice or rats. If developmental effects are detected, control experiments must be performed on the ground and in space, including the use of a space-based 1-g centrifuge, to identify whether gravity or some other element of the space environment induces the developmental abnormalities.

Development of the Vestibular System

Analysis of the development of gravity-sensing systems, including the vestibular system and other systems that interact with it in vertebrates, should be carried out to determine the importance of gravity to their normal development and maintenance. The recommended investigations summarized below should be performed first in ground-based studies to identify appropriate experiments to be performed in space.

- Studies should be performed to define the critical periods for development of the vestibular system. Thus, the critical periods for cellular proliferation, migration, and differentiation and programmed cell death should be identified and the effects of microgravity on these processes assessed.

Neural Space Maps

Neurons composing the brainstem, hippocampal, striatal, and sensory and motor cortical space maps should be investigated as part of the following recommended studies:

- The role of otolithic stimulation on the development and maintenance of the different neural space maps should be investigated.
- Studies should be designed to address how neurons of the various sensory and motor systems interact with vestibular neurons in the normal assembly and function of the neural space maps. Factors should be identified that are supplied by and to the sensory neurons that produce the orderly assembly of these maps in precise coordinate registration.
- The influence of microgravity on the development and maintenance of the space maps should be studied.

Neuroplasticity

It is important to characterize neuroplasticity using multidisciplinary approaches that combine structural and molecular with functional investigations of identified cell populations. The process should be characterized at several different times following the perturbation, in order to determine the sequence of intermediate events leading to the plastic change. Controls for the effects of non-gravitational stresses of types likely to be encountered in space (such as loud noise and vibration) must also be performed on the ground, so that the space-based experiments can be designed to isolate the effects of microgravity from the effects of other stresses. The committee makes the following recommendations for research on neuroplasticity, including one recommendation taken from Chapter 5, "Sensorimotor Integration."

- Studies are needed to determine whether the compensatory mechanisms that normally function in the vestibulomotor pathways are altered by exposure to microgravity. These experiments should be given the highest priority, because these compensatory mechanisms operate in astronauts entering and returning from space and may have a profound effect on their performance in space and their postflight recovery on Earth.
- Experiments are needed to critically test the role of gravity on the development and maintenance of the vestibular system's capability for neuroplasticity.
- Because the vestibulo-oculomotor system is capable of learning new motor patterns in response to sensory perturbations, it is important to determine if and how these mechanisms are affected by exposure to microgravity.
- Functional magnetic resonance imaging (fMRI) should be employed to investigate the following:
    —Changes in sensory and motor cortical maps in human bed-rest studies mimicking different flight durations.

—The effects of microgravity on cortical maps in the human. Pre- and postflight fMRI studies should be conducted with astronauts.

Plants, Gravity, and Space

The study of plants in the space environment has been driven by three main needs: (1) learning how to grow plants successfully in space (space horticulture) either for research or for eventual use in long-term life support systems (2) determining whether there are any plant developmental or metabolic processes that are critically dependent on gravity, and (3) learning how plants alter their patterns of growth and development to respond to changes in the direction of the gravity vector.

Space Horticulture

A major goal of the Advanced Life Support (ALS) program is to develop an effective, completely closed plant growth system capable of growing plants for a bioregenerative life support system. Toward this end, the committee makes the following recommendations:

• The ALS program should concentrate its ground-based research on developing a completely enclosed plant growth system. This effort will require close collaboration between engineers and plant environmental scientists.
• The ALS spaceflight program should focus on testing the potentially gravity-sensitive components of the closed plant growth system, such as the nutrient delivery system.

Role of Gravity in Plant Development

Whether gravity is required for any specific aspect of the development or metabolism of a plant can best be determined by growing a model plant in space through at least two successive generations (seed-to-seed experiment) and examining carefully the development of the resulting plants to ascertain whether any aspect of the development is altered by a lack of gravity. Specifically, the committee recommends the following:

• The seed-to-seed experiment should be the top priority in this area. The promising result obtained with *Brassica rapa* should be confirmed and extended, using *Arabidopsis thaliana* plants. This experiment must be conducted on the ISS, because the plants should be grown through at least two generations in space.
• To conduct a meaningful seed-to-seed experiment, NASA needs to develop the following:
—A superior plant growth unit, with adequate lighting, gas exchange, and water and/or nutrient delivery; and
—*Arabidopsis thaliana* plants that are insensitive to expected environmental stresses and that contain indicator genes for all the expected environmental stresses, such as high levels of CO_2, vibration, anaerobiosis, water stress, and temperature stresses.
• In the interim, before the ISS is functional, studies on specific stages of plant development in space should be limited to small plants with short life cycles (e.g., *Arabidopsis thaliana* or *Brassica rapa*). Whenever possible, a 1-g on-board centrifuge should be available.

Responses of Plants to Change in Direction of the Gravity Vector

Plants respond to the specific direction of the gravity vector in several ways. Among these are the direction of growth of stems and roots (gravitropism) and the swimming direction of some unicellular algae (gravitaxis). Among the committee's recommendations regarding this area of research, the following have the highest priority:

- A primary focus of NASA-sponsored research in plant biology should be on the mechanisms of gravitropism. In particular, modern cellular and molecular techniques should be used to determine the following:
    —The identity of the cells that actually perceive gravity, and the role of the cytoskeleton in the process;
    —The nature of the cellular asymmetry that is set up in a cell that perceives the direction of the gravity vector;
    —The nature and mechanism of the translocation of the signals that pass from the site of perception to the site of reaction; and
    —The nature of the response to the signals that leads to alterations in the rate of cell enlargement.
- A secondary focus should be on the mechanisms of graviperception in single cells, including gravitropic responses of mosses and gravitaxic responses of algae.

Sensorimotor Integration

Sensorimotor integration is an essential element in the control of posture and locomotion, as well as in coordinated body activities such as manipulation of objects and use of tools. The transition from normal gravity to microgravity disrupts postural control and orientation mechanisms. Spatial illusions, and often motion sickness, occur until adaptation to the new force background is achieved. On reentry, severe disturbances of postural, locomotory, and movement control are experienced with reexposure to the normal terrestrial environment. Thresholds for angular and linear accelerations, vestibulo-ocular reflexes, postural mechanisms, vestibulo-spinal reflexes, and gaze control all have been studied extensively in humans, but development of animal models has lagged. Some of these areas require additional study, and a number of new experimental questions arise, given current knowledge and the need to consider human performance during extended-duration space missions.

Spatial Orientation

Future work should emphasize mechanisms related to the active control of body orientation and movement rather than passive thresholds for detection of angular or linear acceleration. Briefly summarized, the committee's research recommendations are as follows:

1. It is of critical interest to determine how microgravity and other unusual force environments, including rotating environments, affect the integrative coordination of eye, head, torso, arm, and leg movements.
2. It is important for the success of long-duration space missions to identify the sensory, motor, and cognitive factors that influence adaptation and retention of adaptation to different force environments, including rotating environments.
3. The influence of altered force levels, including microgravity, on spatial coding of position should be explored in parallel experiments with humans and animals.

Posture and Locomotion

The severe reentry disturbances of posture and locomotion experienced by astronauts and cosmonauts after even short-duration spaceflight pose potentially dangerous operational problems. These disturbances would be especially critical in long-duration missions that require accurate postural, locomotory, and manipulatory control during transitions in background force level. The committee recommends the following:

- The time course for adaptation of locomotion and posture to variations in background force level should be determined.
- Techniques should be developed to provide ancillary sensory inputs or aids to enhance postural and locomotory control during and after transitions between different force levels.

Vestibulo-Ocular Reflexes and Oculomotor Control

Considerable progress has been made in understanding how microgravity affects vestibulo-ocular reflexes, pursuit and saccadic eye movements, and control of gaze. The following studies, which can be carried out in parabolic flight, orbital flight, and rotating rooms, are recommended to achieve closure on understanding these critical functions.

- Systematic parametric studies of pursuit, saccadic, and optokinetic eye movements should be carried out as a function of background force level in humans from microgravity to 2 g.
- The coordination of eye-head-torso synergies in different force levels and their adaptation to changes in force level should be assessed, with the goal of developing a comprehensive three-dimensional model of the vestibulo-ocular reflex and cervical control of gaze.

Space Motion Sickness

Space motion sickness is an operational problem during the first 72 hours of flight, despite the use of medication, and is hazardous for initial transitions between force environments. The use of virtual environment devices in spaceflight to augment training in long-duration missions and for experimental purposes will likely exacerbate motion sickness. Research is recommended on the following:

- The relationship of motion sickness to altered sensorimotor control of head and body in microgravity and greater than 1-g force backgrounds generated in parabolic flight and rotating rooms; and
- The relationship of the vestibular system to autonomic function, especially cardiovascular regulation.

Bone Physiology

One of the best-documented pathophysiological changes associated with microgravity and the spaceflight environment is bone loss, which can exceed 1 percent per month in weight-bearing bones even when an in-flight exercise regime is followed. Within the discipline of bone physiology, the phenomenon of bone loss in astronauts is clearly the issue of greatest concern to NASA. Both the extent and the reversibility of the bone loss are crucial questions for long-term crewed flights on the space station and future space exploration and should be addressed by collecting data from each astronaut to build up the necessary database.

Studies on Humans

The committee recommends that question about microgravity-induced bone loss in humans be studied as follows:

1. To obtain a detailed description of human bone loss in space, a record of skeletal changes, occurring during microgravity and postflight should be generated for each astronaut and correlated with age and

gender, muscle changes, hormonal changes during flight, diet and genetic factors (e.g., susceptibility to osteoporosis) if and when they are known.

2. Bone turnover studies should establish if bone loss is due to increased bone destruction (resorption), decreased bone formation, or both.

3. To develop effective countermeasures, different modalities of mechanical stimulation, the use of exercise (e.g., impact loading), and pharmacological means to prevent bone loss should be evaluated.

Animal Models

If applicable to humans, a considerable amount of useful data on bone loss could be generated using animal models. The committee's priority recommendations are summarized as follows:

1. It should be determined if mechanisms of the bone changes produced by microgravity in animal models are similar to those in humans. Rodent models should include mice, given their smaller size, availability of genetic variants and transgenic animals. Adult animals should be used. In-flight experiments should include animals exposed to centrifugal forces that reproduce 1-g conditions.

2. When an animal model is identified that mimics human bone changes in spaceflight, it should be used in ground-based models of microgravity, such as hindlimb-suspension unloading. If the ground-based model reproduces the changes observed under microgravity conditions, it should be used extensively to address questions of mechanisms.

Skeletal Muscle

A better understanding of the deleterious effects on skeletal muscle of spaceflight and reloading upon return to Earth is necessary to maintain performance and prevent injury. Even after missions of a few weeks, the locomotion of astronauts is very unstable immediately after return to Earth, owing to a combination of orthostatic intolerance, altered otolith-spinal reflexes, reliance on weakened atrophic muscles and inappropriate motor patterns. The committee's high-priority research recommendations are summarized below:

• Priority should be given to research that focuses on cellular and molecular mechanisms underlying muscle weakness, fatigue, incoordination, and delayed-onset muscle soreness.

• Ground-based models, including bed rest for humans and hindlimb unloading in normal and genetically altered rodents, should be used within and across disciplines to investigate the mechanisms underlying in-flight and postflight effects on muscle mass, protein composition, myogenesis, fiber type differentiation, and neuromuscular development.

• The mechanisms should be determined whereby muscle cells sense working length and the mechanical stress of gravity. Signal transduction pathways for growth factors, stretch-activated ion channels, regulators of protein synthesis, and interactions of extracellular matrix and membrane proteins with cytoskeleton should be investigated.

Cardiovascular and Pulmonary Systems

The cardiovascular and pulmonary systems undergo major changes in microgravity, including reduced blood volume that is redistributed headward, increased heart volume, altered blood pressure and heart rate, and improved gas exchange in the lungs despite the surprising persistence of lung ventilation-perfusion inequalities. Many observational research questions have been answered. Future research should focus more on mechanisms. The committee developed a number of recommendations for specific research studies which are broadly summarized below.

Cardiovascular System

• Reevaluate current antiorthostatic countermeasures, and develop and validate new ones. Priority should be given to interventions that may provide simultaneous bone and/or muscle protection.

• Extend current knowledge regarding the magnitude, time course, and mechanisms of cardiovascular adjustments to include long-duration microgravity.

• Determine the mechanisms underlying inadequate total peripheral resistance observed during postflight orthostatic stress.

• Identify and validate appropriate methods for referencing intrathoracic vascular pressures to systemic pressures in microgravity.

Pulmonary System

• Characterize gravity-determined topographical differences of blood flow, ventilation, alveolar size, intrapleural pressures, and mechanical stresses in microgravity during rest and exercise.

• Determine the extent to which pulmonary vascular and microvascular pressures and lymphatic flow are altered by microgravity and whether these changes have any impact on either aging or disease processes.

• Examine patterns of aerosol deposition, and determine whether ventilatory and nonventilatory responses to particulate or antigen inhalation are altered by microgravity.

• Identify changes in pulmonary function that occur during extravehicular activity (EVA), and establish resuscitation procedures for crew members in the event of loss of cabin or EVA suit pressure.

• Evaluate respiratory muscle structure and function in microgravity, at rest, and during maximal exercise.

Endocrinology

The endocrine, nervous, and immune systems regulate the human response to spaceflight and the readjustment processes that follow landing. The principal spaceflight responses to which there is a significant endocrine contribution are the fluid shifts, perturbation of circadian rhythms, loss of red cell mass, possible alterations in the immune system, losses of bone and muscle, and maintenance of energy balance. With the advent of the space station era, the focus shifts from early responses to spaceflight to the long-term adaptive response. The three chronic responses which are areas of serious concern are bone loss, muscle atrophy, and possibly the question of maintaining energy balance at an acceptable level. Priority should be given to studies that are designed to do the following:

• Ensure adequate dietary input during spaceflight. Energy intake must meet needs, and physiologial measurements must be made on subjects in approximate energy balance so that measurements are not confounded by an undernutrition response. The relationship between the amount of exercise and protein and energy balance in flight should be investigated.

• Obtain a human hormone profile early and late in flight and, as a control, preflight measurements on the same individuals over an extended period of time.

• Define the effects of spaceflight on human circadian rhythms. If significant degradation of performance is found and it can be attributed to the disturbed circadian rhythm, explore the use of countermeasures, including a combination of light and melatonin.

Immunology

As individuals stay longer in space, the potential effects of spaceflight on immune function become more significant. There is now convincing evidence that immunological parameters are affected by spaceflight, and important questions should be answered regarding both the biological and the medical significance of these effects and their mechanisms. Future immunological studies should concentrate on functional immunological changes which have been shown to be biologically and medically significant.

Animal Studies

Rodent studies can be used to determine the biological and/or biomedical significance of spaceflight-induced changes in immune responses. Both short- and long-term studies should be carried out, with priority given to those briefly summarized below:

1. Resistance to infection should be examined in animals immediately after their return from spaceflight.
2. Acquired immune responses should be examined, including specific humoral and cellular immune responses.

Human Studies

Immunological measurements and testing of humans should be carried out to examine parameters with potential functional consequences. The recommended studies are briefly summarized below:

1. Acquired immune responses should be examined as described above for animals.
2. Innate immune responses should be examined, including natural killer cell and neutrophil function.
3. Epidemiological studies should be conducted, as the population of astronauts and cosmonauts increases, to assess the potential risk of infection and, in particular, of the development of tumors.

Radiation Hazards

Exposure of crew members to radiation in space poses potentially serious health effects that need to be controlled or mitigated before long-term missions beyond low Earth orbit can be initiated. The levels of radiation in interplanetary space are high enough and the missions long enough that adequate shielding is necessary to minimize carcinogenic, cataractogenic, and possible neurologic effects for crew members.

The knowledge needed to design adequate radiation shielding has both physical and biological components: (1) the distribution and energies of radiation particles present behind a given shielding material as a result of the shield being struck by a given type and level of incident radiation and (2) the effects of a given dose on relevant biological systems for different radiation types. Each component involves significant uncertainty that must be reduced to permit effective design of shielding, given that the level of uncertainty governs the amount of shielding.[6]

The execution of the recommended strategies will require considerably more beam time at a heavy-ion accelerator than is currently available, and it is recommended that NASA explore various possibilities, including the construction of new facilities, to increase the research time available for experiments with high-atomic-number, high-energy (HZE) particles. Priority should be given to the following studies:

1. Determine the carcinogenic risks following irradiation by protons and HZE particles.
2. Determine how cell killing and induction of chromosomal aberrations vary as a function of the thickness and composition of shielding.
3. Determine whether there are studies that can be conducted to increase the confidence of extrapolation from rodents to humans of radiation-induced genetic alterations that in turn could enhance similar extrapolations for cancer.

APPENDIX A

4. Determine if exposure to heavy ions at the level that would occur during deep-space missions of long duration poses a risk to the integrity and function of the central nervous system.

5. Determine if a better error analyses can be performed of all factors contributing to estimation of risk by a particular method, and what are the types and magnitude of uncertainty associated with each method.

6. Determine how the selection and design of the space vehicle affects the radiation environment in which the crew has to exist.

Behavioral Issues

Long-duration missions in space are likely to produce significant changes in individual, group, and organizational behavior. Future missions in space will involve longer periods of exposure to features of the physical environment unique to space and features of the psychosocial environment characteristic of isolated and confined environments. Evidence from previous space missions and from analogue studies suggests that behavioral responses to these environmental stressors will be influenced by characteristics of the individuals, groups, and organizations involved in long-duration missions.

The following list broadly summarizes, in order of priority, the recommended research for behavior and performance during long-duration missions in space:

1. Develop noninvasive qualitative and quantitative techniques for the ongoing assessment of preflight, in-flight, and postflight behavior and performance.

2. Investigate the neurobiological and psychosocial mechanisms underlying the effects of physical and psychosocial environmental stressors on cognitive, affective, and psychophysiological measures of behavior and performance. Such research should be conducted both in space and in ground-based analogue environments.

—Research on environmental factors should include an assessment of affective and cognitive responses to microgravity-related changes in perceptual and physiological systems and behavioral responses to perceived physical dangers, restricted privacy and personal space, and physical and social monotony.

—Research on physiological factors should include studies of behavioral correlates of changes in circadian rhythms and sleep patterns; change and stability in individual physiological patterns in response to psychosocial and environmental stress and their applicability to measures of in-flight behavior and performance; and the relationship between self-reports and external performance-related and physiological symptoms of stress.

—Research on individual factors should include studies of specific coping strategies and behavioral and physiological indicators of coping-stage transitions during long-duration missions; associations between general and mission-specific personality characteristics and performance criteria of ability, stability, and compatibility; changes in problem-solving ability and other aspects of cognitive performance in flight; and changes in personality and behavior postflight.

—Research on interpersonal factors should include studies of the influence of crew psychosocial heterogeneity on crew tension, cohesion, and performance during the mission; factors affecting ground-crew interactions; and the influence of different styles of leadership and decision-making procedures on group performance.

—Research on organizational factors should include studies of the effect of differences in the cultures of the participating agencies on individual and group performance and behavior; the association between mission duration and changes in behavior and performance; and the organizational requirements for effective management of long-duration missions as they relate to task scheduling and workload and the distribution of authority and decision making.

3. Evaluate existing countermeasures and develop new countermeasures that effectively contribute to optimal levels of crew performance, individual well-being, and mission success. These countermeasures include the following:

—Screening and selection procedures that are based on a "select-in" assessment of individual personality characteristics and interpersonally oriented psychological assessments of crew compatibility;

—Training programs that are team oriented and that enable crews to successfully address the social, cultural, and psychological issues likely to occur in flight;

—Organizational countermeasures for filling unstructured time and reducing boredom and monotony;

—Clinical countermeasures, such as the use of psychoactive medications in microgravity environments and the use of voice analysis for monitoring the interpersonal performance of crews; and

—Design of spacecraft interiors and amenities to maximize control over the physical environment and reduce impacts of physical monotony on behavior and performance.

CROSSCUTTING RESEARCH PRIORITIES

This section summarizes the committee's recommendations for the highest-priority research across the entire spectrum of space life sciences. In the near term, until the research facilities of the International Space Station come online or an additional Spacelab mission is provided, NASA-supported research will necessarily be directed primarily to ground-based investigations designed to answer fundamental questions and frame critical hypotheses that can later be tested in space. Indeed, as this report emphasizes, understanding the basic mechanisms underlying biological and behavioral responses to spaceflight is essential to designing effective countermeasures and protecting astronaut health and safety both in space and upon return to Earth. For these reasons, the following recommendations for high-priority areas of crosscutting research place emphasis on ground-based studies.

Physiological and Behavioral Effects of Spaceflight

Priority should be given to research aimed at ameliorating problems that may limit astronauts' health, safety, or performance during and after long-term spaceflight. The committee emphasizes that specific priorities may shift to a significant degree depending on the types of missions to be carried out in the future, particularly as related to long-term human exploration of space. For this reason, the recommended areas of research are not given a priority order.

Loss of Weight-bearing Bone and Muscle

Bone loss and muscle deterioration are among the best-documented deleterious effects caused by spaceflight in humans and animals. Exercise has been only partially successful in preventing muscle weakness and bone loss. Development of effective countermeasures requires advances in several areas of research.

- Research should emphasize studies that provide mechanistic insights into development of effective countermeasures for preventing bone and muscle deterioration during and after spaceflight.
- Ground-based model systems, such as hindlimb unloading of rodents, should be used to investigate the mechanisms of changes that reproduce in-flight and postflight effects.
- A database on the course of microgravity-related bone loss and its reversibility in humans should be established in preflight, in-flight, and postflight recording of bone mineral density.
- Hormonal profiles should be obtained on humans before, during, and after spaceflight.
- The relationship between exercise activity levels and protein energy balance in flight should be investigated.

Vestibular Function, the Vestibular Ocular Reflex, and Sensorimotor Integration

During the transitions in gravitational force that occur going into and returning from spaceflight, the vestibular system undergoes changes in activity that can result in debilitating symptoms in astronauts.

• The highest priority should be given to studies designed to determine the bases for the adaptive compensatory mechanisms in the vestibular and sensorimotor systems that operate both on ground and in space.
• In-flight recordings of signal processing following otolith afferent stimulation should be made to determine how exposure to microgravity affects central and peripheral vestibular function and development.
• Motor learning should be investigated in spaceflight and the results compared with findings obtained in ground-based studies of this process.

Orthostatic Intolerance Upon Return to Earth Gravity

Orthostatic hypotension, present since the very earliest human spaceflights, still affects a high percentage of astronauts returning from spaceflights even of relatively short duration and is an even greater problem for shuttle pilots, who must perform complex reentry maneuvers in an upright, seated position. The problem remains despite the use of extensive antiorthostatic countermeasures by both U.S. and Russian space programs. Studies should focus on determining physiological mechanisms and developing effective countermeasures.

• Current knowledge of the magnitude, time course, and mechanisms of cardiovascular adjustments should be extended to include long-duration exposure to microgravity.
• The specific mechanisms underlying inadequate total peripheral resistance observed during postflight orthostatic stress should be determined.
• Current antiorthostatic countermeasures should be reevaluated to refine those that offer protection and eliminate those that do not. Priority should be given to interventions that may provide simultaneous bone and/or muscle protection.
• Appropriate methods for referencing intrathoracic vascular pressures to systemic pressures in microgravity should be identified and validated, given the observed changes in cardiac and pulmonary volume and compliance.

Radiation Hazards

The biological effects of exposure to radiation in space pose potentially serious health effects for crew members in long-term missions beyond low Earth orbit. High priority is given to the following recommended studies:

• Determine the carcinogenic risks following irradiation by protons and high-atomic-number, high-energy (HZE) particles.
• Determine if exposure to heavy ions at the level that would occur during deep-space missions of long duration poses a risk to the integrity and function of the central nervous system.
• Determine how the selection and design of the space vehicle affect the radiation environment in which the crew has to exist.
• Determine whether combined effects of radiation and stress on the immune system in spaceflight could produce additive or synergistic effects on host defenses.

Physiological Effects of Stress

The immune system interacts closely with the neuroendocrine system. Results indicate a close association between the neuroendocrine status of the host and host defense systems.

- The role that the host response to stressors during spaceflight plays in alterations in host defenses should be determined.

Psychological and Social Issues

The health, well-being, and performance of astronauts on extended missions may be negatively affected by many stressful aspects of the space environment. Mechanisms of response to physiological and psychosocial stressors encountered in spaceflight must be better understood in order to ensure crew safety, health, and productivity.

- Highest priority should be given to interdisciplinary research on the neurobiological (circadian, endocrine) and psychosocial (individual, group, organizational) mechanisms underlying the effects of physical and psychosocial environmental stressors. Cognitive, affective, and psychophysiological measures of behavior and performance should be examined in ground-based analogue settings as well as in flight.
- High priority should be given to evaluation of existing countermeasures (screening and selection, training, monitoring, support) and development of effective new countermeasures.

Fundamental Gravitational Biology

Mechanisms of Graviperception and Gravitropism in Plants

Plants respond to changes in the direction of the gravitational vector by altering the direction of growth of roots and stems. The gravitropic response requires (1) perception of the gravitational vector by gravisensing cells; (2) intracellular transduction of this information; (3) translocation of the resulting signal to the sites of reaction, i.e., sites of differential growth; and (4) reaction to the signal by the responding cells, i.e., initiation of differential growth.

- Studies on graviperception should concentrate on three problems:
  —The identity of the cells that actually perceive gravity;
  —The intracellular mechanisms by which the direction of the gravity vector is perceived; and
  —The threshold value for graviperception—this will require a spaceflight experiment.
- Studies on gravitropic transduction should focus on the nature of the cellular asymmetry that is set up in a cell that perceives the direction of the gravity vector.
- Studies on the translocation step should concentration on the nature and mechanism of the translocation of the signals that pass from the site of perception to the site of reaction..
- Studies on the reaction step should focus on the mechanism(s) by which the gravitropic signals cause unequal rates of cell elongation, and on possible effects of gravity on the sensitivity of these cells to the signals.

Mechanisms of Graviperception in Animals

It is known that in several systems sensory stimulation plays a role in the development of the neural connections necessary for normal processing of sensory information. The potential role of gravity in the normal development of the gravity-sensing vestibular system of animals is therefore an important area for ground- and space-based research.

- Ground-based studies should identify the critical periods in vestibular neuron development before initiation of experiments on the effects of microgravity on vestibular development.
- Pre- and postflight functional magnetic resonance imaging (fMRI) studies should be conducted with astronauts to determine the effects of microgravity on neural space maps.

Effects of Spaceflight on Reproduction and Development

To determine whether there are developmental processes that are critically dependent on gravity, organisms should be grown through at least two full generations in space.

- Key model animals should be grown through two life cycles; the highest priority should be given to vertebrate models. If significant developmental effects are detected, control experiments must be performed to determine whether gravity or some other element of the space environment induces these developmental abnormalities.
- An analogous experiment should be carried out with the model plant *Arabidopsis thaliana* to confirm results obtained on Mir with a preliminary experiment using *Brassica rapa*.

PROGRAMMATIC AND POLICY ISSUES

Although NASA has responded effectively to many of the programmatic and policy issues raised in the 1987 and 1991 reports,[7,8] significant concerns in the program and policy arena remain unresolved. These focus on issues relating to strategic planning and conduct of space-based research; utilization of the International Space Station (ISS) for life sciences research; mechanisms for promoting integrated and interdisciplinary research; collection of and access to human flight data, specifically; publication of and access to space life sciences research in general; and professional education.

Space-based Research

Development of Advanced Instrumentation and Methodologies

Future life science flight experiments on ISS will depend on the availability of advanced instrumentation to carry out the measurements and analyses required by the research questions and approaches described in this report. In addition, facile data and information transfer between space- and ground-based investigators is crucial.

- NASA should work with the broad life sciences community to identify and catalyze the development of advanced instrumentation and methodologies that will be required for sophisticated space-based research in the coming decade.
- NASA should take advantage of advanced instrumentation developed in other countries.
- The capability of direct, real-time communication between space-based experimenters and principal investigators at their home laboratories should be a high-priority objective for the ISS.

Utilization of International Space Station for Life Sciences Research

Issues relating to design and use of the ISS are a major concern of the committee. These issues include (1) changes in design of ISS, (2) diversion of funds for scientific facilities and equipment into construction budgets, (3) the adequacy of power and data transmission to and from Earth, (4) the availability of crew time for research, and (5) an extended hiatus in flight opportunities for life sciences research owing to delays in ISS construction. These issues have alarmed the life sciences communities.

- To better ensure that ISS will adequately meet the needs of space life sciences researchers, NASA should continue to bring the external user community as well as NASA scientists into the planning and design phases of facility construction.
- NASA should make every effort to mount at least one Spacelab life sciences flight in the period between Neurolab and the completion of ISS facilities.
- NASA should determine whether continuation of shuttle missions for short-term flight experiments after the opening of ISS would be economically and scientifically sound.

Science Policy Issues

Peer Review

The Division of Life Sciences initiated a universal system of peer review in 1994 for all NASA-supported investigators. The new process has the committee's strong support.

- Responsibility for the establishment of peer review panels and for funding decisions should remain a function of the Headquarters Division of Life Sciences.
- NASA should regularly evaluate the composition of scientific review panels to ensure that the feasibility of proposed flight experiments receives appropriate expert evaluation.

Integration of Research Activities

- Principal investigators of projected flight experiments should be brought together with NASA managers and design engineers at the beginning of the planning process to function as an integrated team responsible for all phases of the planning, design and testing. This integration should continue throughout the life of the project.
- NASA should regularly review and evaluate the NASA Specialized Centers of Research and Training (NSCORT) program to determine whether this mechanism provides the best way to foster interdisciplinary research and increase the scientific value of the life sciences research program.
- NASA should regularly review and evaluate the performance of the National Space Biomedical Research Institute and the impact of its funding on the overall life sciences research budget and program.

Human Flight Data: Collection and Access

The disciplinary chapters of this report repeatedly stress the need for improved, systematic collection of data on astronauts preflight, in space, and postflight.

- NASA should initiate an ISS-based program to collect detailed physiological and psychological data on astronauts before, during, and after flight.
- NASA should make every effort to promote mechanisms for making complete data obtained from studies on astronauts accessible to qualified investigators in a timely manner. Consideration should be given to possible modifications of current policies and practices relating to confidentiality of human subjects that would ethically ensure astronaut cooperation in a more effective manner.

Publication and Outreach

An essential outcome of scientific research is publication—dissemination of results to the scientific community at large. The record of peer-reviewed publication, especially of spaceflight experiments, by funded investigators in NASA's life sciences programs needs to be improved, as does the usefulness of the Spaceline Archive to the scientific community.

- NASA should provide funding for data analysis and publication of flight experiments for a sufficient period to insure analysis of the data and publication of the results.
- NASA should insist on timely dissemination of the results of space life sciences research in peer-reviewed publications. For investigators with previous NASA support, the publication record should be an important criterion for subsequent funding.
- NASA should take as a high priority the completion of data entry into the Spaceline Archive and should ensure that access to the archive is simple and transparent.

Professional Education

NASA should make every effort to ensure the professional training of graduate students and postdoctoral fellows in space and gravitational biology and medicine.

- NASA should take as high priority the support of a small, highly competitive program of postdoctoral fellowships for training in laboratories of NASA-supported investigators in academic and research institutions external to NASA centers.

REFERENCES

1. Space Science Board, National Research Council. 1987. A Strategy for Space Biology and Medical Science for the 1980s and 1990s. National Academy Press, Washington, D.C.
2. Space Science Board, 1987, A Strategy for Space Biology and Medical Science for the 1980s and 1990s, p. xi.
3. Space Science Board, 1987, A Strategy for Space Biology and Medical Science for the 1980s and 1990s, p. 4.
4. Space Studies Board, National Research Council. 1991. Assessment of Programs in Space Biology and Medicine 1991. National Academy Press, Washington, D.C.
5. Space Science Board, National Research Council. 1987. A Strategy for Space Biology and Medical Science for the 1980s and 1990s. National Academy Press, Washington, D.C.
6. Wilson, J.W., Cucinotta, F.A., Shinn, J.L., Kim, M.H., and Badavi, F.F. 1997. Shielding strategies for human space exploration: Introduction. Chapter 1 in Shielding Strategies for Human Space Exploration: A Workshop (John W. Wilson, Jack Miller, and Andrei Konradi, eds.). National Aeronautics and Space Administration.
7. Space Science Board, National Research Council. 1987. A Strategy for Space Biology and Medical Science for the 1980s and 1990s. National Academy Press, Washington, D.C.
8. Space Studies Board, National Research Council. 1991. Assessment of Programs in Space Biology and Medicine 1991. National Academy Press, Washington, D.C.

Appendix B

Letter of Request from NASA

APPENDIX B

National Aeronautics and
Space Administration

Headquarters
Washington, DC 20546-0001

Reply to Attn of: UL

OCT 15 1998

Dr. Claude R. Canizares
Space Studies Board, HA 584
National Academy of Sciences
2101 Constitution Avenue, NW
Washington, DC 20418

Dear Dr. Canizares:

The Committee on Space Biology and Medicine recently completed a major report, *A Strategy for Research in Space Biology and Medicine in the New Century*, which reviewed the status of space life sciences research in all of the disciplines funded by NASA's life sciences program, and laid out a comprehensive strategy for research during the next decade. In that report, numerous biomedical research questions related to astronaut health and safety were identified as critical to NASA's long-duration flight program. I would like to request that the committee now undertake an assessment of NASA's current program in biomedical research in light of the recommendations of the *Strategy* report.

Specifically, we would ask that the Committee on Space Biology and Medicine (CDBM) undertake a review of NASA's entire biomedical research program, both intramural and extramural, and assess the degree to which the program meets research needs over the next ten years. The research priorities given in the recent CSBM report, *A Strategy for Research in Space Biology and Medicine in the New Century*, could be used as a point of departure when considering future needs and priorities. It would be helpful if the committee would examine the relationship between intramural and extramural biomedical research activities sponsored by the agency, and review the content and program organization of both. The roles of the NASA Specialized Centers of Research and Training and the National Space Biomedical Research Institute, in the biomedical program, could also be examined. Such a review should cover all NASA biomedical research activities at NASA, including those currently conducted in conjunction with operational medical and aerospace medicine programs.

Some of the specific items which I would hope to see considered in the report are:

- The balance of discipline areas emphasized in the current program.

- The degree to which studies of fundamental cellular and physiological mechanisms are addressed in each discipline program.

- The balance between ground and flight investigations.

- NASA plans for the development and validation of physiological and psychological countermeasures.

- Plans for validation of animal models.

- Extent to which programs are supporting new, advanced technologies and methodologies.

In order to carry out such a study, the committee will no doubt find it necessary to visit make site visits to the NASA centers responsible for directing or carrying out biomedical research. We understand that completion of the study would require approximately two years from its inception, so that delivery of a report in the fall of 2000 might be reasonably expected.

Sincerely,

Joan Vernikos, Ph.D.
Director, Life Sciences Division

Appendix C

Glossary

adrenergic agents: Compounds with actions similar to those of the adrenal hormones epinephrine (adrenaline) and norepinephrine that stimulate the sympathetic nervous system.

arrhythmia: Alteration in the rhythm of the heartbeat either in time or in force.

autonomic regulation: Regulation of the portion of the nervous system that modulates blood pressure, heart rate, and other involuntary functions.

baroreceptor: Receptor that responds to changes in blood pressure.

Bion: Russian space capsule that can support animals (e.g., monkeys, rats) and insects in orbit up to three weeks.

bone mineral density: Bone mineral content corrected for bone size (bone mineral content/cross-sectional area).

bone remodeling: Continuous process of breakdown and renewal of bone that occurs throughout life.

chromosome translocation: Transfer of a segment of one chromosome to another nonhomologous chromosome.

colony-stimulating factor: Protein that causes division and maturation of immature bone marrow cells into mature white blood cells.

corpus striatum: Neuroanatomical term for cerebral gray matter in three basal ganglia, including the caudate nucleus, putamen, and globus pallidus; involved in the control of movement.

countermeasure: Any step taken to prevent or alleviate decrements in the physiological or psychological status of astronauts that may result from spaceflight. Measures may include in-flight protocols such as medication and exercise, as well as the screening of astronaut candidates using medical selection criteria.

cytokine: Class of proteins that mediates immune (and other) responses.

dosimetry: Experimental procedure used to determine dose (joules per kilogram) of ionizing radiation.

functional magnetic resonance imaging (fMRI): Noninvasive scanning technology used to produce images of localized neural activity in the human brain. It is based on the fact that there are changes in local blood flow and blood oxygenation in response to neural activity.

hindlimb unloading: Simulation of some aspects of microgravity effects on rodents by unloading of muscles and head-down tilt by employing various harnessing strategies to elevate the animal and prevent its hindlimbs from weight bearing on the floor.

histomorphometric: Referring to structural changes in tissues observed and quantitated on biopsy specimens.

hypothalamus-pituitary-adrenal (HPA) axis: Neuroendocrine axis that interacts with the immune response; mediates response to stress and other factors that could have a profound effect on immune response and resistance.

leukocyte: White blood cell.

leukocyte distribution: Distribution of subtypes of white blood cells (e.g., CD4+ and CD8+ lymphocyte distribution); indication of host potential for normal immune function.

leukocyte proliferation: Response of lymphocytes and macrophages (white blood cells) to stimulation in which lymphocytes divide after receiving signals from macrophages; evidence of intact immune function.

locomotor: Affecting or involving the organs that mediate active movements.

natural killer cells: Immune cells found in the body under normal conditions that can destroy target virus-infected and tumor cells by an as-yet-unknown recognition mechanism.

neuroendocrine: Referring to the complex integration of the central nervous system with pituitary hormone secretion.

Neurolab: Shuttle mission (1998) dedicated to studies in neurosciences.

neuroplasticity: Ability of differentiated neurons to alter structure or function in response to altered physiological demands.

neurovestibular: Pertaining to the sensory system concerned with maintenance of posture and balance by perceiving gravity (linear acceleration) and rotary movements of the head.

oculomotor: Related to the control of eye movement.

orthostatic hypotension: Drop in blood pressure when going from a lying or sitting position to a standing position.

orthostatic intolerance: Inability to stand erect without a drop in blood pressure that induces weakness or fainting.

osteoblast: Bone-forming cell.

osteoclast: Multinucleated, bone-resorbing cell related to macrophages.

otoconia: Biogenerated calcium carbonate crystals embedded in a gelatinous membrane accelerated during body movements and stimulating receptor hair cells of the inner ear.

otolith: Otoconial matrix mass suspended on hair cells of the inner ear sensitive to linear acceleration.

proprioceptive: Conscious awareness of positions of various parts of the body in space provided by joint and muscle sensory inputs; also, sensory stimulation arising from receptors within muscles.

sensorimotor: Combined sensory and neural aspects of neural-dependent body activity.

skin test reaction: Reaction due to cell-mediated immune processes on skin when individuals are exposed to substances to which they have been exposed previously (e.g., tubercle bacillus).

space map: Organization of groups of neurons into specific arrangements in the central nervous system that reflect the structure and/or function of the spatial environment representing the sensory or motor areas of the brain.

space motion sickness: Condition similar to terrestrial motion sickness, often encountered by astronauts on entry into orbit; typical symptoms include pallor, increased body warmth, cold sweating, dizziness, nausea, and vomiting.

teleoperation: Remote control by a human operator of a robot or machine such as an extraterrestrial exploration vehicle.

telepresence: State of a human operator feeling physically present at a remote site because of the degree of realism of the information transmitted about the machine being controlled remotely.

thermoregulation: Process by which mammals control body temperature.

translocation yields: Fraction of cells containing chromosome translocations, usually per 1,000 cells observed.

vestibular: Pertaining to the sensory system concerned with maintenance of posture and balance by perceiving gravity (linear acceleration) and rotary movements of the head.

vestibulo-collic: Neuronal connections formed by axons of vestibular nuclei neurons with motor neurons in cervical spinal cord, which activate neck muscles to support the head during head movements.

vestibulo-ocular (or vestibular ocular) reflex: Passive eye movements elicited by activation of receptors in the vestibular apparatus.

Appendix D

Acronyms

AGS: Alternating Gradient Synchrotron

ALARA: as low as reasonably achievable

AMERD: Astronaut Medical Evaluation Requirements Document

ARC: Ames Research Center (NASA)

BAF: Booster Application Facility

BP: Behavior and Performance

BR&C: Biomedical Research and Countermeasures

CNS: central nervous system

CSBM: Committee on Space Biology and Medicine

CV: cardiovascular

DSO: Detailed Supplemental Objective

DXA: dual energy x-ray absorptiometry

EDOMP: Extended Duration Orbiter Medical Project

EKG: electrocardiogram

ESA: European Space Agency

EVA: extravehicular activity

FBRP: Fundamental Biology Research Program, formerly Gravitational Biology and Ecology (GB&E)

fMRI: functional magnetic resonance imaging

GB&E: Gravitational Biology and Ecology, recently renamed the Fundamental Biology Research Program (FBRP)

GH: growth hormone

HEDS: Human Exploration and Development of Space (NASA)

HPA: hypothalamus-pituitary-adrenal

HZE: high atomic number, high energy

IGF-I: insulin-like growth factor-I

IPT: integrated product team

ISS: International Space Station

ITR: Integrated Testing Regimen

JSC: Johnson Space Center (NASA)

LBNP: lower-body negative pressure

LEO: low Earth orbit

LET: linear energy transfer

MARES: Muscle Atrophy Research and Exercise System

MORD: Medical Operations Requirements Document

MRI: magnetic resonance imaging

NASA: National Aeronautics and Space Administration

NIH: National Institutes of Health

NRA: NASA Research Announcement

NRC: National Research Council

NSBRI: National Space Biomedical Research Institute

NSCORT: NASA Specialized Centers of Research and Training

PI: principal investigator

PRN: as needed

RBE: relative biological effectiveness

SHFE: Space Human Factors Engineering

SMSP: Shuttle Mir Space Program

SPE: solar particle event

USRA: Universities Space Research Association

VOR: vestibulo-ocular reflex

Appendix E

Biographies of Committee Members

Mary Jane Osborn, *Chair,* is professor and head of microbiology at the University of Connecticut Health Center. Dr. Osborn's specialties are biochemistry, microbiology, and molecular biology, and her current research interests include the biogenesis of bacterial membranes. Dr. Osborn has served on numerous distinguished committees, including the National Science Board (1980-1986), the President's Committee on the National Medal of Sciences (1981-1982), the Advisory Council of the National Institutes of Health's Division of Research Grants (1989-1994; chair, 1992-1994), the Advisory Council of the Max Planck Institute of Immunobiology (1974-1978), the Board of Scientific Advisors for the Roche Institute for Molecular Biology (1981-1985; chair, 1983-1985), and the Governing Board of the National Research Council (1990-1993). Dr. Osborn is a member of the National Academy of Sciences, the American Association for the Advancement of Science, the American Society of Biochemistry and Molecular Biology (President, 1981-1982), the American Chemical Society (chair, Division of Biological Chemistry, 1975-1976), the American Academy of Arts and Sciences (fellow; Council, 1988-1992), the Federation of American Societies for Experimental Biology (president, 1982-1983), the American Society for Microbiololgy, and the American Academy of Microbiology.

Norma M. Allewell is associate vice president for sponsored programs and technology transfer at Harvard University. She has expertise in the fields of molecular biophysics, structural biology, and biochemistry, and her research interests include protein structure, function, and design; macromolecular interactions; and computer modeling. Dr. Allewell is a member of the Biophysical Society (president, 1993-1994), the American Association for the Advancement of Science (fellow), the American Society for Biochemistry and Molecular Biology, and Sigma Xi.

Jay C. Buckey, Jr, is a research associate professor of medicine in the Department of Medicine at the Dartmouth-Hitchcock Medical School and staff physician at the White River Junction Veteran's Administration Medical Center. He was coinvestigator on cardiovascular adaptation experiments on the SLS-1 space shuttle mission and, more recently, was payload specialist astronaut on the Neurolab space mission, STS-90, where the experiments focused on the effects of microgravity on the brain and nervous system.

Lynette Jones is a principal research scientist in the Department of Mechanical Engineering and Research Laboratory of Electronics at the Massachusetts Institute of Technology. Her primary research is on the human proprioceptive system and the role of muscle and cutaneous mechanoreceptors in sensory processes. This research has led to studies of haptic interfaces that are used to interact with computer-generated virtual environments and teleoperated robots. She also does research on the development of wearable health monitoring devices, and she is involved in developing a portable system to evaluate the visual-vestibular system and a non-invasive method to measure glucose levels in people with diabetes. Dr. Jones is a member of the International Society for Psychophysics and the Society for Neuroscience.

Robert A. Marcus is director of a program in clinical disorders of bone and mineral metabolism at Stanford University. His primary research interests are acquisition, maintenance, and regulation of bone mass in humans. His laboratory studies hormonal nutrition and physical activity determinants of bone mass.

Lawrence A. Palinkas is a professor in the Division of International Health and Cross-Cultural Medicine, Department of Family and Preventive Medicine at the University of California, San Diego. He is also the director of the UCSD Immigrant/Refugee Health Studies Program and a faculty member of the UCSD/San Diego State University Joint Doctoral Program in Clinical Psychology. A medical anthropologist with expertise in behavioral and cross-cultural medicine, he counts as his primary research interests behavior and performance in isolated and confined extreme environments and the cultural context of stress, coping, and illness. He has served on numerous NASA and U.S. Navy advisory groups on behavior and performance. He is a member of the American Public Health Association, the American Anthropological Association (fellow), the Society for Medical Anthropology, the Aerospace Medical Association, the Society for Behavioral Medicine, and the American Psychosomatic Society.

Kenna D. Peusner is a professor of anatomy and cell biology at the George Washington University School of Medicine. Dr. Peusner is a neurobiologist specializing in intracellular electrophysiological and microscopic techniques to investigate neural structure and function. Her research is focused on characterizing synaptic transmission and ionic conductances and their role in the emergence of excitability in the developing and damaged central vestibular system. She is a member of the Neuroscience Society, the Association for Research in Otolaryngology, the New York Academy of Science, and the American Association for the Advancement of Science. Dr. Peusner received the Lindback Foundation award for distinguished teaching in the basic medical sciences, Jefferson Medical College. She is a grantee of the National Institute on Deafness and Other Communicative Disorders, National Institutes of Health.

Steven E. Pfeiffer is a professor of microbiology at the University of Connecticut Medical School. He has expertise in molecular cell biology and neurobiology, and his research interests are in molecular, cell, and developmental biology of the nervous system and myelinogenesis. Organizations of which he is a member include the American Association of Cell Biologists, the American Society for Neurochemistry, the International Society for Developmental Biology (president, 1996-1998), the International Society for Neurochemistry, and the Society for Neuroscience.

Danny A. Riley is a professor of anatomy and cell biology at the Medical College of Wisconsin. Dr. Riley's expertise is in the mechanisms of muscle atrophy and nerve regeneration in animal models and humans, with an emphasis on space biology. He was a recipient of the American Institute of Aviation and Astronautics Jeffries Medical Research Award in 1992 for outstanding contributions to the advancement of aerospace medical research and two NASA Group Achievement awards—for the Cosmos 2044 Biosatellite Team (91) and the Spacelab Life Sciences-2 Team (93). He is an elected member of the Board of Directors of the American Society for Gravitational and Space Biology (1989-1993, 1997-present). His other memberships include the American Association of Anatomists, the International Society of Electromyographic Kinesiology, the Society for Neuroscience, the American Society for Cell Biology, the Aerospace Medical Association, and the American Institute of Biological Sciences.

Richard Setlow is associate director for life sciences at Brookhaven National Laboratory. Dr. Setlow is an expert in the fields of radiation biophysics and molecular biology and his research interests include far ultraviolet spectroscopy; ionizing and nonionizing radiation; molecular biophysics; action of light on proteins viruses and cells; nucleic acids; repair mechanisms; and environmental carcinogenesis. Dr. Setlow received the Finsen Medal in 1980 for "outstanding contribution to photobiology and repair of nucleic acids" and the Enrico Fermi Award in 1989 from the U.S. Department of Energy for "pioneering and far-reaching contributions to the fields of radiation biophysics and molecular biology." His memberships include the National Academy of Sciences, the American Association for the Advancement of Science, the Biophysical Society, the American Society for Photobiology, the Environmental Mutagen Society, and the American Association for Cancer Research.

Gerald Sonnenfeld is professor and chair of the Department of Microbiology and Immunology and associate dean for basic sciences and graduate studies at the Morehouse School of Medicine. His expertise in the field of immunology is in interferon and cytokine research. Dr. Sonnenfeld has served on numerous peer review and advisory groups for NASA's and other agencies' immunology research programs and as program director of NASA's Space Biology Research Associates Program. Dr. Sonnenfeld is past president of the American Society for Gravitational and Space Biology. His other memberships include the American Association of Immunologists, the American Society for Microbiology, the American Society for Virology, the International Cytokine Society, the International Society for Interferon Research (charter member), the International Society for Antiviral Research, Sigma Xi, the Society for Leukocyte Biology, and the Tissue Engineering Society (founding board member).

T. Peter Stein is a professor of surgery and nutrition at the University of Medicine and Dentistry of New Jersey. His expertise is in the areas of clinical nutrition and protein and energy metabolism during spacelift; lipid metabolism; clinical nutrition; nutritional assessment; and lung biochemistry. Dr. Stein was a co-winner of the American Institute of Aviation and Astronautics Jeffries Medical Research Award in 1992 for his work on Spacelab Life Sciences-1. His memberships include the American Association for the Advancement of Science, the American Institute of Nutrition, the American Society for Clinical Nutrition, the American Physiological Society, the Society for Parenatal and Enteral Nutrition, the American Chemical Society, the American College of Nutrition, and the American Society for Gravitational Physiology.

Judith L. Swain is chair of the Department of Medicine and professor of medicine at Stanford University. Before joining the staff at Stanford, Dr. Swain was the Herbert C. Rorer Professor of Medical Science, professor of genetics, and director of cardiovascular medicine at the University of Pennsylvania. Her research expertise includes the study of cardiovascular disease, cardiovascular developmental biology, and angiogenesis. Dr. Swain is a member of the Institute of Medicine.